Selected Titles in This Series

Volume

7 **S. R. S. Varadhan**
  Probability theory
  2001

6 **Louis Nirenberg**
  Topics in nonlinear functional analysis
  2001

5 **Emmanuel Hebey**
  Nonlinear analysis on manifolds: Sobolev spaces and inequalities
  2000

3 **Percy Deift**
  Orthogonal polynomials and random matrices: A Riemann-Hilbert approach
  2000

2 **Jalal Shatah and Michael Struwe**
  Geometric wave equations
  2000

1 **Qing Han and Fanghua Lin**
  Elliptic partial differential equations
  2000

Courant Lecture Notes in Mathematics

Executive Editor
Jalal Shatah

Managing Editor
Paul D. Monsour

Assistant Editor
Reeva Goldsmith

Copy Editor
Will Klump

S. R. S. Varadhan
Courant Institute of Mathematical Sciences

7 Probability Theory

Courant Institute of Mathematical Sciences
New York University
New York, New York

American Mathematical Society
Providence, Rhode Island

2000 *Mathematics Subject Classification.* Primary 60–01.

Library of Congress Cataloging-in-Publication Data
Varadhan, S. R. S.
   Probability theory / S.R.S. Varadhan.
     p. cm. — (Courant lecture notes ; 7)
   Includes bibliographical references and index.
   ISBN 0-8218-2852-5 (alk. paper)
   1. Probabilities. I. Title. II. Series.

QA273.V348 2001
519.2—dc21   2001045216

      Copying and reprinting. Individual readers of this publication, and nonprofit libraries acting for them, are permitted to make fair use of the material, such as to copy a chapter for use in teaching or research. Permission is granted to quote brief passages from this publication in reviews, provided the customary acknowledgment of the source is given.
      Republication, systematic copying, or multiple reproduction of any material in this publication is permitted only under license from the American Mathematical Society. Requests for such permission should be addressed to the Assistant to the Publisher, American Mathematical Society, P. O. Box 6248, Providence, Rhode Island 02940-6248. Requests can also be made by e-mail to reprint-permission@ams.org.

                              © 2001 by the author. All rights reserved.
                  The American Mathematical Society retains all rights
                  except those granted to the United States Government.
                          Printed in the United States of America.
      ∞ The paper used in this book is acid-free and falls within the guidelines
                    established to ensure permanence and durability.
                Visit the AMS home page at URL: http://www.ams.org/
                        10 9 8 7 6 5 4 3 2 1   06 05 04 03 02 01

Contents

Preface vii

Chapter 1. Measure Theory 1
   1.1. Introduction 1
   1.2. Construction of Measures 3
   1.3. Integration 7
   1.4. Transformations 13
   1.5. Product Spaces 14
   1.6. Distributions and Expectations 16

Chapter 2. Weak Convergence 19
   2.1. Characteristic Functions 19
   2.2. Moment-Generating Functions 22
   2.3. Weak Convergence   *Most important section!* 24

Chapter 3. Independent Sums 35
   3.1. Independence and Convolution 35
   3.2. Weak Law of Large Numbers 37
   3.3. Strong Limit Theorems 40
   3.4. Series of Independent Random Variables 43
   3.5. Strong Law of Large Numbers 48
   3.6. Central Limit Theorem 49
   3.7. Accompanying Laws 54
   3.8. Infinitely Divisible Distributions 59
   3.9. Laws of the Iterated Logarithm 66

Chapter 4. Dependent Random Variables 73
   4.1. Conditioning 73
   4.2. Conditional Expectation 79
   4.3. Conditional Probability 81
   4.4. Markov Chains 84
   4.5. Stopping Times and Renewal Times 89
   4.6. Countable State Space 90
   4.7. Some Examples 98

Chapter 5. Martingales 109
   5.1. Definitions and Properties   *Lemma 5.4!* 109
   5.2. Martingale Convergence Theorems 112

5.3.	Doob Decomposition Theorem	115
5.4.	Stopping Times	117
5.5.	Up-crossing Inequality	120
5.6.	Martingale Transforms, Option Pricing	121
5.7.	Martingales and Markov Chains	123

Chapter 6. Stationary Stochastic Processes   131

6.1.	Ergodic Theorems	131
6.2.	Structure of Stationary Measures	135
6.3.	Stationary Markov Processes	137
6.4.	Mixing Properties of Markov Processes	141
6.5.	Central Limit Theorem for Martingales	143
6.6.	Stationary Gaussian Processes	147

Chapter 7. Dynamic Programming and Filtering   157

7.1.	Optimal Control	157
7.2.	Optimal Stopping	158
7.3.	Filtering	161

Bibliography   163

Index   165

Preface

These notes are based on a first-year graduate course on probability and limit theorems given at Courant Institute of Mathematical Sciences. Originally written during the academic year 1996-97, they have been subsequently revised during the academic year 1998-99 as well as in the Fall of 1999. Several people have helped me with suggestions and corrections and I wish to express my gratitude to them. In particular, I want to mention Prof. Charles Newman, Mr. Enrique Loubet, and Ms. Vera Peshchansky. Chuck used it while teaching the same course in 1998–99, Enrique helped me as TA when I taught from these notes again in the fall of 1999, and Vera, who took the course in the fall of 2000, provided me with a detailed list of corrections. These notes cover about three-fourths of the course, essentially discrete time processes. Hopefully there will appear a companion volume some time in the near future that will cover continuous time processes. A small amount of measure theory is included. While it is not meant to be complete, it is my hope that it will be useful.

CHAPTER 1

Measure Theory

1.1. Introduction

The evolution of probability theory was based more on intuition rather than mathematical axioms during its early development. In 1933, A. N. Kolmogorov [4] provided an axiomatic basis for probability theory, and it is now the universally accepted model. There are certain "noncommutative" versions that have their origins in quantum mechanics (see, for instance, K. R. Parthasarathy [5]) that are generalizations of the Kolmogorov model. We shall, however, use Kolmogorov's framework exclusively.

The basic intuition in probability theory is the notion of randomness. There are experiments whose results are not predictable and can be determined only after performing them and then observing the outcomes. The simplest familiar examples are the tossing of a fair coin and the throwing of a balanced die. In the first experiment the result could be either a head or a tail, and in the second a score of any integer from 1 through 6. These are experiments with only a finite number of alternate outcomes. It is not difficult to imagine experiments that have countably or even uncountably many alternatives as possible outcomes.

Abstractly then, there is a space Ω of all possible outcomes and each individual outcome is represented as a point ω in that space Ω. Subsets of Ω are called *events* and each of them corresponds to a collection of outcomes. If the outcome ω is in the subset A, then the event A is said to have occurred. For example, in the case of a die the set $A = \{1, 3, 5\} \subset \Omega$ corresponds to the event "an odd number shows up." With this terminology it is clear that the union of sets corresponds to "or," intersection to "and," and complementation to "negation."

One would expect that probabilities should be associated with each outcome and there should be a *probability function* $f(\omega)$ which is the probability that ω occurs. In the case of coin tossing we may expect $\Omega = \{H, T\}$ and

$$f(T) = f(H) = \frac{1}{2},$$

or in the case of a die

$$f(1) = f(2) = \cdots = f(6) = \frac{1}{6}.$$

Since probability is normalized so that certainty corresponds to a probability of 1, one expects

(1.1) $$\sum_{\omega \in \Omega} f(\omega) = 1.$$

If Ω is uncountable this is a mess. There is no reasonable way of adding up an uncountable set of numbers, each of which is 0. This suggests that it may not be possible to start with probabilities associated with individual outcomes and build a meaningful theory. The next best thing is to start with the notion that probabilities are already defined for events. In such a case, $P(A)$ is defined for a class $\mathcal{B}$ of subsets $A \subset \Omega$. The question that arises naturally is what should $\mathcal{B}$ be and what properties should $P(\cdot)$ defined on $\mathcal{B}$ have? It is natural to demand that the class $\mathcal{B}$ of sets for which probabilities are to be defined satisfy the following properties:

The whole space Ω and the empty set $\emptyset$ are in $\mathcal{B}$. For any two sets A and B in $\mathcal{B}$, the sets $A \cup B$ and $A \cap B$ are again in $\mathcal{B}$. If $A \in \mathcal{B}$, then the complement A^c is again in $\mathcal{B}$. Any class of sets satisfying these properties is called a *field*. (finite ops!)

DEFINITION 1.1 A *probability*, or more precisely a *finitely additive probability measure*, is a nonnegative set function $P(\cdot)$ defined for sets $A \in \mathcal{B}$ that satisfies the following properties:

(1.2) $$P(A) \geq 0 \quad \text{for all } A \in \mathcal{B},$$

(1.3) $$P(\Omega) = 1 \quad \text{and} \quad P(\emptyset) = 0.$$

If $A, B \in \mathcal{B}$ are disjoint, then

(1.4) $$P(A \cup B) = P(A) + P(B).$$

In particular,

(1.5) $$P(A^c) = 1 - P(A)$$

for all $A \in \mathcal{B}$.

P on a field

A condition which is somewhat more technical, but important from a mathematical viewpoint, is that of countable additivity. The class $\mathcal{B}$, in addition to being a field, is assumed to be closed under countable unions (or equivalently, countable intersections); i.e., if $A_n \in \mathcal{B}$ for every n, then $A = \bigcup_n A_n \in \mathcal{B}$. Such a class is called a *σ-field*. The "probability" itself is presumed to be defined on a σ-field $\mathcal{B}$.

DEFINITION 1.2 A set function P defined on a σ-field is called a *countably additive probability measure* if in addition to satisfying equations (1.2), (1.3), and (1.4), it satisfies the following countable additivity property: for any sequence of pairwise *disjoint* sets A_n with $A = \bigcup_n A_n$

P on a σ-field

(1.6) $$P(A) = \sum_n P(A_n).$$

EXERCISE 1.1. The limit of an increasing (or decreasing) sequence A_n of sets is defined as its union $\bigcup_n A_n$ (or the intersection $\bigcap_n A_n$). A monotone class is defined as a class that is closed under monotone limits of an increasing or decreasing sequence of sets. Show that a field $\mathcal{B}$ is a σ-field if and only if it is a monotone class. ⇒ obvious, ⇐ make sets like a sq

monotone class

EXERCISE 1.2. Show that a finitely additive probability measure $P(\cdot)$ defined on a σ-field $\mathcal{B}$ is countably additive, i.e., satisfies equation (1.6), if and only if it satisfies either of the following two equivalent conditions:

⇐ is clear by making redefining sets
⇒ obvious

1.2. CONSTRUCTION OF MEASURES

- If A_n is any nonincreasing sequence of sets in $\mathcal{B}$ and $A = \lim_n A_n = \bigcap_n A_n$, then
$$P(A) = \lim_n P(A_n).$$

- If A_n is any nondecreasing sequence of sets in $\mathcal{B}$ and $A = \lim_n A_n = \bigcup_n A_n$, then
$$P(A) = \lim_n P(A_n).$$

EXERCISE 1.3. If $A, B \in \mathcal{B}$ and P is a finitely additive probability measure, show that $P(A \cup B) = P(A) + P(B) - P(A \cap B)$. How does this generalize to $P(\bigcup_{j=1}^n A_j)$?

EXERCISE 1.4. If P is a finitely additive measure on a field $\mathcal{F}$ and $A, B \in \mathcal{F}$, show that $|P(A) - P(B)| \leq P(A \triangle B)$ where $A \triangle B$ is the symmetric difference $(A \cap B^c) \cup (A^c \cap B)$. In particular, if $B \subset A$,
$$0 \leq P(A) - P(B) = P(A \cap B^c) \leq P(B^c).$$

EXERCISE 1.5. If P is a countably additive probability measure, show that for any sequence $A_n \in \mathcal{B}$, $P(\bigcup_{n=1}^\infty A_n) \leq \sum_{n=1}^\infty P(A_n)$.

Although we would like our "probability" to be a countably additive probability measure, on a σ-field $\mathcal{B}$ of subsets of a space Ω it is not clear that there are plenty of such things. As a first small step, show the following:

EXERCISE 1.6. If $\{\omega_n : n \geq 1\}$ are distinct points in Ω and $p_n \geq 0$ are numbers with $\sum_n p_n = 1$, then
$$P(A) = \sum_{n:\omega_n \in A} p_n$$
defines a countably additive probability measure on the σ-field of all subsets of Ω. (This is still cheating because the measure P lives on a countable set.)

DEFINITION 1.3 A probability measure P on a field $\mathcal{F}$ is said to be *countably additive* on $\mathcal{F}$ if for any sequence $A_n \in \mathcal{F}$ with $A_n \downarrow \emptyset$, we have $P(A_n) \downarrow 0$.

EXERCISE 1.7. Given any class $\mathcal{F}$ of subsets of Ω there is a unique σ-field $\mathcal{B}$ such that it is the smallest σ-field that contains $\mathcal{F}$.
Hint. Take $\mathcal{B} = \bigcap_{\Sigma \in \mathcal{A}} \Sigma$ where
$$\mathcal{A} = \{\Sigma : \Sigma \supset \mathcal{F}, \Sigma \text{ is a } \sigma\text{-field}\}.$$

DEFINITION 1.4 The σ-field in the above exercise is called the *σ-field generated by $\mathcal{F}$*.

1.2. Construction of Measures

The following theorem is important for the construction of countably additive probability measures. A detailed proof of this theorem, as well as other results on measure and integration, can be found in [3, 7] or in any one of the many texts on real variables. In an effort to be complete, we will sketch the standard proof.

1. MEASURE THEORY

THEOREM 1.1 (Caratheodory Extension Theorem) *Any countably additive probability measure P on a field $\mathcal{F}$ extends uniquely as a countably additive probability measure to the σ-field $\mathcal{B}$ generated by $\mathcal{F}$.*

PROOF: The proof proceeds along the following steps:
Step 1. Define an object P^* called the *outer measure* for all sets A.

$$(1.7) \qquad P^*(A) = \inf_{\cup_j A_j \supset A} \sum_j P(A_j)$$

where the infimum is taken over all countable collections $\{A_j\}$ of sets from $\mathcal{F}$ that cover A. Without loss of generality we can assume that $\{A_j\}$ are disjoint. (Replace A_j by $(\cap_{i=1}^{j-1} A_i^c) \cap A_j$).

Step 2. Show that P^* has the following properties:

(1) P^* is countably *subadditive*, i.e.,

$$P^*\left(\bigcup_j A_j\right) \leq \sum_j P^*(A_j).$$

(2) For $A \in \mathcal{F}$, $P^*(A) \leq P(A)$. (Trivial) ✓
(3) For $A \in \mathcal{F}$, $P^*(A) \geq P(A)$. (Need to use the countable additivity of P on $\mathcal{F}$).

Step 3. Define a set E to be *measurable* if

$$P^*(A) \geq P^*(A \cap E) + P^*(A \cap E^c)$$

holds for all sets A, and establish the following properties for the class $\mathcal{M}$ of measurable sets: The class of measurable sets $\mathcal{M}$ is a σ-field, and P^* is a countably additive measure on it.

Step 4. Finally, show that $\mathcal{M} \supset \mathcal{F}$. This implies that $\mathcal{M} \supset \mathcal{B}$ and P^* is an extension of P from $\mathcal{F}$ to $\mathcal{B}$.

Uniqueness is quite simple. Let P_1 and P_2 be two countably additive probability measures on a σ-field $\mathcal{B}$. Let us define $\mathcal{A} = \{A : P_1(A) = P_2(A)\}$. Then $\mathcal{A}$ is a monotone class, i.e., if $A_n \in \mathcal{A}$ is increasing (decreasing), then $\cup_n A_n (\cap_n A_n) \in \mathcal{A}$. According to the exercise below, the monotone class $\mathcal{A}$, if it contains the field $\mathcal{F}$, must necessarily contain the σ-field $\mathcal{B}$ generated by $\mathcal{F}$. □

EXERCISE 1.8. The smallest monotone class generated by a field is the same as the σ-field generated by the field.

It now follows that $\mathcal{A}$ must contain the σ-field generated by $\mathcal{F}$ and that proves uniqueness.

The extension theorem does not quite solve the problem of constructing countably additive probability measures. It reduces it to constructing them on fields. The following theorem is important in the theory of Lebesgue integrals and is very useful for the construction of countably additive probability measures on the real line. The proof will again only be sketched. The natural σ-field on which to define a probability measure on the line is the Borel σ-field. This is defined as the smallest σ-field containing all intervals and includes in particular all open sets.

Let us consider the class of subsets of the real numbers, $\mathcal{I} = \{I_{a,b} : -\infty \leq a < b \leq \infty\}$ where $I_{a,b} = \{x : a < x \leq b\}$ if $b < \infty$, and $I_{a,\infty} = \{x : a < x < \infty\}$. In other words $\mathcal{I}$ is the collection of intervals that are left-open and right-closed. The class of sets that are finite-disjoint unions of members of $\mathcal{I}$ is a field $\mathcal{F}$, if the empty set is added to the class. If we are given a function $F(x)$ on the real line which is nondecreasing and satisfies

$$\lim_{x \to -\infty} F(x) = 0 \quad \text{and} \quad \lim_{x \to \infty} F(x) = 1,$$

we can define a finitely additive probability measure P by first defining

$$P(I_{a,b}) = F(b) - F(a)$$

for intervals and then extending it to $\mathcal{F}$ by defining it as the sum for disjoint unions from $\mathcal{I}$. Let us note that the Borel σ-field $\mathcal{B}$ on the real line is the σ-field generated by $\mathcal{F}$.

THEOREM 1.2 (Lebesgue) P is countably additive on $\mathcal{F}$ if and only if $F(x)$ is a right continuous function of x. Therefore for each right continuous nondecreasing function $F(x)$ with $F(-\infty) = 0$ and $F(\infty) = 1$ there is a unique probability measure P on the Borel subsets of the line such that $F(x) = P(I_{-\infty,x})$. Conversely, every countably additive probability measure P on the Borel subsets of the line comes from some F. The correspondence between P and F is one-to-one.

PROOF: The only difficult part is to establish the countable additivity of P on $\mathcal{F}$ from the right continuity of $F(\cdot)$. Let $A_j \in \mathcal{F}$ and $A_j \downarrow \emptyset$, the empty set. Let us assume $P(A_j) \geq \delta > 0$ for all j and then establish a contradiction.

Step 1. We take a large interval $[-\ell, \ell]$ and replace A_j by $B_j = A_j \cap [-\ell, \ell]$. Since $|P(A_j) - P(B_j)| \leq 1 - F(\ell) + F(-\ell)$, we can make the choice of ℓ large enough that $P(B_j) \geq \frac{\delta}{2}$. In other words we can assume without loss of generality that $P(A_j) \geq \frac{\delta}{2}$ and $A_j \subset [-\ell, \ell]$ for some fixed $\ell < \infty$.

Step 2. If

$$A_j = \bigcup_{i=1}^{k_j} I_{a_{j,i}, b_{j,i}}$$

use the right continuity of F to replace A_j by B_j, which is again a union of left-open right-closed intervals with the same right endpoints, but with left endpoints moved ever so slightly to the right. Achieve this in such a way that

$$P(A_j - B_j) \leq \frac{\delta}{10 \cdot 2^j} \quad \text{for all } j.$$

1. MEASURE THEORY

Step 3. Define C_j to be the closure of B_j obtained by adding to it the left endpoints of the intervals making up B_j. Let $E_j = \bigcap_{i=1}^{j} B_i$ and $D_j = \bigcap_{i=1}^{j} C_i$. Then,

(1) the sequence D_j of sets is decreasing,
(2) each D_j is a closed bounded set, and
(3) since $A_j \supset D_j$ and $A_j \downarrow \emptyset$, it follows that $D_j \downarrow \emptyset$.

Because $D_j \supset E_j$ and $P(E_j) \geq \frac{\delta}{2} - \sum_j P(A_j - B_j) \geq \frac{4}{10\delta}$ each D_j is nonempty and this violates the finite intersection property that every decreasing sequence of bounded nonempty closed sets on the real line has a nonempty intersection, i.e., has at least one common point.

The rest of the proof is left as an exercise. □

The function F is called the *distribution function* corresponding to the probability measure P.

EXAMPLE 1.1. Suppose $x_1, x_2, \ldots, x_n, \ldots$ is a sequence of points and we have probabilities p_n at these points, then for the discrete measure

$$P(A) = \sum_{n: x_n \in A} p_n$$

we have the distribution function

$$F(x) = \sum_{n: x_n \leq x} p_n$$

that only increases by jumps, the jump at x_n being p_n. The points $\{x_n\}$ themselves can be discrete like integers or dense like the rationals.

EXAMPLE 1.2. If $f(x)$ is a nonnegative integrable function with integral 1, i.e., $\int_{-\infty}^{\infty} f(y) dy = 1$, then $F(x) = \int_{-\infty}^{x} f(y) dy$ is a distribution function which is continuous. In this case f is the density of the measure P and can be calculated as $f(x) = F'(x)$. a.e.

There are (messy) examples of F that are continuous, but do not come from any density. More on this later.

EXERCISE 1.9. Let us try to construct the Lebesgue measure on the space $\mathcal{Q} \subset [0, 1]$ that consists only of the rationals.
We would like to have

$$P[\mathbb{1}_{a,b}] = b - a$$

for all rational $0 \leq a \leq b \leq 1$. Show that it is impossible by showing that $P[\{q\}] = 0$ for the set $\{q\}$ containing the single rational q while $P[\mathcal{Q}] = P[\bigcup_{q \in \mathcal{Q}} \{q\}] = 1$. Where does the earlier proof break down?

Once we have a countably additive probability measure P on a space (Ω, Σ), we will call the triple (Ω, Σ, P) a *probability space*.

1.3. Integration

An important notion is that of a random variable or a measurable function.

DEFINITION 1.5 A *random variable* or *measurable function* is a map $f : \Omega \to \mathbb{R}$, i.e., a real-valued function $f(\omega)$ on Ω such that for every Borel set $B \subset \mathbb{R}$, $f^{-1}(B) = \{\omega : f(\omega) \in B\}$ is a measurable subset of Ω, i.e., $f^{-1}(B) \in \Sigma$.

EXERCISE 1.10. It is enough to check the requirement for sets $B \subset \mathbb{R}$ that are intervals or even just sets of the form $(-\infty, x]$ for $-\infty < x < \infty$.

A function that is measurable and satisfies $|f(\omega)| \leq M$ for all $\omega \in \Omega$ for some finite M is called a bounded measurable function.

The following statements are the essential steps in developing an integration theory. Details can be found in any one of several books on real variables including [3] and [5].

(1) If $A \in \Sigma$, the indicator function of A, defined as

$$\mathbf{1}_A(\omega) = \begin{cases} 1 & \text{if } \omega \in A \\ 0 & \text{if } \omega \notin A \end{cases}$$

is bounded and measurable.

(2) Sums, products, limits, compositions, and reasonable elementary operations like min and max performed on measurable functions lead to measurable functions.

(3) If $\{A_j : 1 \leq j \leq n\}$ is a finite disjoint partition of Ω into measurable sets, the function $f(\omega) = \sum_j c_j \mathbf{1}_{A_j}(\omega)$ is a measurable function and is referred to as a "simple" function.

(4) Any bounded measurable function f is a uniform limit of simple functions. To see this, if f is bounded by M, divide $[-M, M]$ into n non-overlapping subintervals I_j of length $\frac{2M}{n}$ with midpoints c_j. Let

$$A_j = f^{-1}(I_j) = \{\omega : f(\omega) \in I_j\} \quad \text{and} \quad f_n = \sum_{j=1}^n c_j \mathbf{1}_{A_j}.$$

Clearly f_n is simple, $\sup_\omega |f_n(\omega) - f(\omega)| \leq \frac{M}{n}$, and we are done.

(5) For simple functions $f = \sum c_j \mathbf{1}_{A_j}$ the integral $\int f(\omega) dP$ is defined to be $\sum_j c_j P(A_j)$. It enjoys the following properties:

   (a) If f and g are simple, so is any linear combination $af + bg$ for real constants a and b and

$$\int (af + bg) dP = a \int f dP + b \int g dP.$$

   (b) If f is simple so is $|f|$ and $|\int f dP| \leq \int |f| dP \leq \sup_\omega |f(\omega)|$.

(6) If f_n is a sequence of simple functions converging to f uniformly, then $a_n = \int f_n dP$ is a Cauchy sequence of real numbers and therefore has a limit a as $n \to \infty$. The integral $\int f dP$ of f is defined to be this limit

a. One can verify that *a depends only on f and not on the sequence f_n chosen to approximate f*.

(7) Now the integral is defined for all bounded measurable functions and enjoys the following properties:

(a) If f and g are bounded measurable functions and a and b are real constants, then the linear combination $af + bg$ is again a bounded measurable function, and

$$\int (af + bg)dP = a \int f\,dP + b \int g\,dP.$$

(b) If f is a bounded measurable function, so is $|f|$ and $|\int f\,dP| \leq \int |f|\,dP \leq \sup_\omega |f(\omega)|$.

(c) In fact, a slightly stronger inequality is true. For any bounded measurable f,

$$\int |f|\,dP \leq P(\{\omega : |f(\omega)| > 0\}) \sup_\omega |f(\omega)|.$$

(d) If f is a bounded measurable function and A is a measurable set, one defines

$$\int_A f(\omega)\,dP = \int \mathbf{1}_A(\omega) f(\omega)\,dP$$

and we can write for any measurable set A,

$$\int f\,dP = \int_A f\,dP + \int_{A^c} f\,dP.$$

In addition to uniform convergence, there are other weaker notions of convergence.

DEFINITION 1.6 A sequence of functions f_n is said to *converge to a function f everywhere* or *pointwise* if

$$\lim_{n\to\infty} f_n(\omega) = f(\omega) \quad \text{for every } \omega \in \Omega.$$

In dealing with sequences of functions on a space that has a measure defined on it, often it does not matter if the sequence fails to converge on a set of points that is insignificant. For example, if we are dealing with the Lebesgue measure on the interval $[0, 1]$ and $f_n(x) = x^n$, then $f_n(x) \to 0$ for all x except $x = 1$. A single point, being an interval of length 0, should be insignificant for the Lebesgue measure.

DEFINITION 1.7 A sequence f_n of measurable functions is said to *converge to a measurable function f almost everywhere* or *almost surely* (usually abbreviated as a.e.) if there exists a measurable set N with $P(N) = 0$ such that

$$\lim_{n\to\infty} f_n(\omega) = f(\omega) \quad \text{for every } \omega \in N^c.$$

1.3. INTEGRATION

Note that almost everywhere convergence is always relative to a probability measure.

Another notion of convergence is the following:

DEFINITION 1.8 A sequence f_n of measurable functions is said to *converge to a measurable function f in measure* or *in probability* if

$$\lim_{n\to\infty} P[\omega : |f_n(\omega) - f(\omega)| \geq \varepsilon] = 0 \quad \text{for every } \varepsilon > 0.$$

Let us examine these notions in the context of indicator functions of sets $f_n(\omega) = \mathbf{1}_{A_n}(\omega)$. As soon as $A \neq B$, $\sup_\omega |\mathbf{1}_A(\omega) - \mathbf{1}_B(\omega)| = 1$, so that uniform convergence never really takes place. On the other hand, one can verify that $\mathbf{1}_{A_n}(\omega) \to \mathbf{1}_A(\omega)$ for every ω if and only if the two sets

$$\limsup_n A_n = \bigcap_n \bigcup_{m \geq n} A_m \quad \text{and} \quad \liminf_n A_n = \bigcup_n \bigcap_{m \geq n} A_m$$

both coincide with A. Finally, $\mathbf{1}_{A_n}(\omega) \to \mathbf{1}_A(\omega)$ in measure if and only if

$$\lim_{n\to\infty} P(A_n \Delta A) = 0$$

where for any two sets A and B the symmetric difference $A \Delta B$ is defined as $A \Delta B = (A \cap B^c) \cup (A^c \cap B) = A \cup B \cap (A \cap B)^c$. It is the set of points that belong to either set but not to both. For instance, $\mathbf{1}_{A_n} \to 0$ in measure if and only if $P(A_n) \to 0$.

EXERCISE 1.11. There is a difference between almost everywhere convergence and convergence in measure. The first is really stronger. Consider the interval $[0, 1]$ and divide it successively into $2, 3, 4, \ldots$, parts and enumerate the intervals in succession. That is, $I_1 = [0, \frac{1}{2}]$, $I_2 = (\frac{1}{2}, 1]$, $I_3 = [0, \frac{1}{3}]$, $I_4 = (\frac{1}{3}, \frac{2}{3}]$, $I_5 = (\frac{2}{3}, 1]$, and so on. If $f_n(x) = \mathbf{1}_{I_n}(x)$, it easy to check that f_n tends to 0 in measure but not almost everywhere.

EXERCISE 1.12. But the following statement is true. If $f_n \to f$ as $n \to \infty$ in measure, then there is a subsequence f_{n_j} such that $f_{n_j} \to f$ almost everywhere as $j \to \infty$.

EXERCISE 1.13. If $\{A_n\}$ is a sequence of measurable sets, then in order that $\limsup_{n\to\infty} A_n = \emptyset$, it is necessary and sufficient that

$$\lim_{n\to\infty} P\left[\bigcup_{m=n}^{\infty} A_m\right] = 0.$$

In particular, it is sufficient that $\sum_n P[A_n] < \infty$. Is it necessary?

LEMMA 1.3 If $f_n \to f$ almost everywhere, then $f_n \to f$ in measure.

PROOF: $f_n \to f$ outside N is equivalent to

$$\bigcap_n \bigcup_{m \geq n} [\omega : |f_m(\omega) - f(\omega)| \geq \varepsilon] \subset N$$

for every $\varepsilon > 0$. In particular, by countable additivity

$$P[\omega : |f_n(\omega) - f(\omega)| \geq \varepsilon] \leq P\left[\bigcup_{m \geq n} [\omega : |f_m(\omega) - f(\omega)| \geq \varepsilon]\right] \to 0$$

as $n \to \infty$

and we are done. $\square$

EXERCISE 1.14. Countable additivity is important for this result. On a finitely additive probability space it could be that $f_n \to f$ everywhere and still $f_n \not\to f$ in measure. In fact, show that if every sequence f_n that converges to 0 everywhere also converges in probability to 0, then the measure is countably additive.

THEOREM 1.4 (Bounded Convergence Theorem) *If the sequence $\{f_n\}$ of measurable functions is uniformly bounded and if $f_n \to f$ in measure as $n \to \infty$, then $\lim_{n \to \infty} \int f_n \, dP = \int f \, dP$.*

PROOF: Since

$$\left|\int f_n \, dP - \int f \, dP\right| = \left|\int (f_n - f) \, dP\right| \leq \int |f_n - f| \, dP$$

we need only prove that if $f_n \to 0$ in measure and $|f_n| \leq M$, then $\int |f_n| \, dP \to 0$. To see this

$$\int |f_n| \, dP = \int_{|f_n| \leq \varepsilon} |f_n| \, dP + \int_{|f_n| > \varepsilon} |f_n| \, dP \leq \varepsilon + MP[\omega : |f_n(\omega)| > \varepsilon]$$

and taking limits

$$\limsup_{n \to \infty} \int |f_n| \, dP \leq \varepsilon$$

and since $\varepsilon > 0$ is arbitrary we are done. $\square$

The bounded convergence theorem is the essence of countable additivity. Let us look at the example of $f_n(x) = x^n$ on $0 \leq x \leq 1$ with Lebesgue measure. Clearly $f_n(x) \to 0$ a.e. and therefore in measure. While the convergence is not uniform, $0 \leq x^n \leq 1$ for all n and x and so the bounded convergence theorem applies. In fact,

$$\int_0^1 x^n \, dx = \frac{1}{n+1} \to 0.$$

However, if we replace x^n by nx^n, $f_n(x)$ still goes to 0 a.e., but the sequence is no longer uniformly bounded and the integral does not go to 0.

We now proceed to define integrals of nonnegative measurable functions.

DEFINITION 1.9 *If f is a nonnegative measurable function we define*

$$\int f \, dP = \left\{\sup \int g \, dP : g \text{ bounded}, 0 \leq g \leq f\right\}.$$

An important result is

1.3. INTEGRATION

THEOREM 1.5 (Fatou's Lemma) *If for each $n \geq 1$, $f_n \geq 0$ is measurable and $f_n \to f$ in measure as $n \to \infty$, then*

$$\int f \, dP \leq \liminf_{n \to \infty} \int f_n \, dP.$$

PROOF: Let us suppose that g is bounded and satisfies $0 \leq g \leq f$. Then the sequence $h_n = f_n \wedge g = \min(f_n, g)$ is uniformly bounded and

$$h_n \to h = f \wedge g = g.$$

Therefore, by the bounded convergence theorem,

$$\int g \, dP = \lim_{n \to \infty} \int h_n \, dP.$$

Since $\int h_n \, dP \leq \int f_n \, dP$ for every n it follows that

$$\int g \, dP \leq \liminf_{n \to \infty} \int f_n \, dP.$$

As g satisfying $0 \leq g \leq f$ is arbitrary, we are done. (by taking sup over g) □

COROLLARY 1.6 (Monotone Convergence Theorem) *If for a sequence $\{f_n\}$ of nonnegative functions we have $f_n \uparrow f$ monotonically, then*

$$\int f_n \, dP \to \int f \, dP \quad \text{as } n \to \infty.$$

PROOF: Obviously $\int f_n \, dP \leq \int f \, dP$ and the other half follows from Fatou's lemma. □

Now we try to define integrals of arbitrary measurable functions. A nonnegative measurable function is said to be *integrable* if $\int f \, dP < \infty$. A measurable function f is said to be integrable if $|f|$ is integrable and we define $\int f \, dP = \int f^+ \, dP - \int f^- \, dP$ where $f^+ = f \vee 0$ and $f^- = -f \wedge 0$ are the positive and negative parts of f. The integral has the following properties:

(1) It is linear. If f and g are integrable, so is $af + bg$ for any two real constants and $\int (af + bg) \, dP = a \int f \, dP + b \int g \, dP$.
(2) $|\int f \, dP| \leq \int |f| \, dP$ for every integrable f.
(3) If $f = 0$ except on a set N of measure 0, then f is integrable and $\int f \, dP = 0$. In particular, if $f = g$ almost everywhere, then $\int f \, dP = \int g \, dP$. approximate by simples to f^+

THEOREM 1.7 (Jensen's Inequality) *If $\phi(x)$ is a convex function of x and $f(\omega)$ and $\phi(f(\omega))$ are integrable, then*

(1.8) $$\int \phi(f(\omega)) \, dP \geq \phi \left(\int f(\omega) \, dP \right).$$

1. MEASURE THEORY

PROOF: We have seen the inequality already for $\phi(x) = |x|^p$. The proof is quite simple. We note that any convex function ϕ can be represented as the supremum of a collection of affine linear functions,

$$(1.9) \qquad \phi(x) = \sup_{(a,b) \in E} \{ax + b\}.$$

It is clear that if $(a, b) \in E$, then $af(\omega) + b \leq \phi(f(\omega))$ and on integration this yields $am + b \leq E[\phi(f(\omega))]$ where $m = E[f(\omega)]$. Since this is true for every $(a, b) \in E$, in view of the representation (1.9), our theorem follows. $\square$

Another important theorem is

THEOREM 1.8 (Dominated Convergence Theorem) *If for some sequence $\{f_n\}$ of measurable functions we have $f_n \to f$ in measure and $|f_n(\omega)| \leq g(\omega)$ for all n and ω for some integrable function g, then $\int f_n \, dP \to \int f \, dP$ as $n \to \infty$.*

PROOF: $g + f_n$ and $g - f_n$ are nonnegative and converge in measure to $g + f$ and $g - f$, respectively. By Fatou's lemma

$$\liminf_{n \to \infty} \int (g + f_n) dP \geq \int (g + f) dP.$$

Since $\int g \, dP$ is finite we can subtract it from both sides and get

$$\liminf_{n \to \infty} \int f_n \, dP \geq \int f \, dP.$$

Working the same way with $g - f_n$ yields

$$\limsup_{n \to \infty} \int f_n \, dP \leq \int f \, dP.$$

and we are done. $\square$

EXERCISE 1.15. Take the unit interval with the Lebesgue measure and define $f_n(x) = n^\alpha \mathbf{1}_{[0, 1/n]}(x)$. Clearly $f_n(x) \to 0$ for $x \neq 0$. On the other hand, $\int f_n(x) dx = n^{\alpha - 1} \to 0$ if and only if $\alpha < 1$. What is $g(x) = \sup_n f_n(x)$ and when is g integrable?

If $h(\omega) = f(\omega) + ig(\omega)$ is a complex-valued measurable function with real and imaginary parts $f(\omega)$ and $g(\omega)$ that are integrable, we define

$$\int h(\omega) dP = \int f(\omega) dP + i \int g(\omega) dP.$$

EXERCISE 1.16. Show that for any complex function $h(\omega) = f(\omega) + ig(\omega)$ with measurable f and g, $|h(\omega)|$ is integrable if and only if $|f|$ and $|g|$ are integrable, and we then have

$$\left| \int h(\omega) dP \right| \leq \int |h(\omega)| dP.$$

1.4. Transformations

A **measurable** space $(\Omega, \mathcal{B})$ is a set Ω together with a σ-field $\mathcal{B}$ of subsets of Ω.

DEFINITION 1.10 Given two measurable spaces $(\Omega_1, \mathcal{B}_1)$ and $(\Omega_2, \mathcal{B}_2)$, a mapping or a transformation $T: \Omega_1 \to \Omega_2$, i.e., a function $\omega_2 = T(\omega_1)$ that assigns for each point $\omega_1 \in \Omega_1$ a point $\omega_2 = T(\omega_1) \in \Omega_2$, is said to be **measurable** if for every measurable set $A \in \mathcal{B}_2$, the inverse image

$$T^{-1}(A) = \{\omega_1 : T(\omega_1) \in A\} \in \mathcal{B}_1.$$

EXERCISE 1.17. Show that, in the above definition, it is enough to verify the property for $A \in \mathcal{A}$ where $\mathcal{A}$ is any class of sets that generates the σ-field $\mathcal{B}_2$.

If T is a measurable map from $(\Omega_1, \mathcal{B}_1)$ into $(\Omega_2, \mathcal{B}_2)$ and P is a probability measure on $(\Omega_1, \mathcal{B}_1)$, the *induced* probability measure Q on $(\Omega_2, \mathcal{B}_2)$ is defined by

$$(1.10) \qquad Q(A) = P(T^{-1}(A)) \quad \text{for } A \in \mathcal{B}_2.$$

EXERCISE 1.18. Verify that Q indeed does define a probability measure on $(\Omega_2, \mathcal{B}_2)$.

The induced measure Q is denoted by PT^{-1}.

THEOREM 1.9 *If $f: \Omega_2 \to \mathbb{R}$ is a real-valued measurable function on Ω_2, then $g(\omega_1) = f(T(\omega_1))$ is a measurable real-valued function on $(\Omega_1, \mathcal{B}_1)$. Moreover, g is integrable with respect to P if and only if f is integrable with respect to Q and*

$$(1.11) \qquad \int_{\Omega_2} f(\omega_2) dQ = \int_{\Omega_1} g(\omega_1) dP.$$

PROOF: If $f(\omega_2) = \mathbf{1}_A(\omega_2)$ is the indicator function of a set $A \in \mathcal{B}_2$, the claim in equation (1.11) is in fact, the definition of measurability and the induced measure. We see, by linearity, that the claim extends easily from indicator functions to simple functions. By uniform limits, the claim can now be extended to bounded measurable functions. Monotone limits then extend it to nonnegative functions. By considering the positive and negative parts separately, we are done. □

A measurable transformation is just a generalization of the concept of a random variable introduced in Section 1.2. We can either think of a random variable as a special case of a measurable transformation where the target space is the real line or think of a measurable transformation as a random variable with values in an arbitrary target space. The induced measure $Q = PT^{-1}$ is called the *distribution* of the random variable T under P. In particular, if T takes real values, Q is a probability distribution on $\mathbb{R}$.

EXERCISE 1.19. When T is real-valued, show that

$$\int T(\omega) dP = \int x \, dQ.$$

When $F = (f_1, f_2, \ldots, f_n)$ takes values in $\mathbb{R}^n$, the induced distribution Q on R^n is called the *joint distribution* of the n random variables $f_1, f_2, \ldots, f_n$.

EXERCISE 1.20. If T_1 is a measurable map from $(\Omega_1, \mathcal{B}_1)$ into $(\Omega_2, \mathcal{B}_2)$ and T_2 is a measurable map from $(\Omega_2, \mathcal{B}_2)$ into $(\Omega_3, \mathcal{B}_3)$, then show that $T = T_2 \circ T_1$ is a measurable map from $(\Omega_1, \mathcal{B}_1)$ into $(\Omega_3, \mathcal{B}_3)$. If P is a probability measure on $(\Omega_1, \mathcal{B}_1)$, then on $(\Omega_3, \mathcal{B}_3)$ the two measures PT^{-1} and $(PT_1^{-1})T_2^{-1}$ are identical.

1.5. Product Spaces

Given two sets Ω_1 and Ω_2, the Cartesian product $\Omega = \Omega_1 \times \Omega_2$ is the set of pairs (ω_1, ω_2) with $\omega_1 \in \Omega_1$ and $\omega_2 \in \Omega_2$. If Ω_1 and Ω_2 come with σ-fields $\mathcal{B}_1$ and $\mathcal{B}_2$, respectively, we can define a natural σ-field $\mathcal{B}$ on Ω as the σ-field generated by sets (measurable rectangles) of the form $A_1 \times A_2$ with $A_1 \in \mathcal{B}_1$ and $A_2 \in \mathcal{B}_2$. This σ-field will be called the product σ-field.

EXERCISE 1.21. Show that sets that are finite-disjoint unions of measurable rectangles constitute a field $\mathcal{F}$.

DEFINITION 1.11 The *product σ-field* $\mathcal{B}$ is the σ-field generated by the field $\mathcal{F}$.

Given two probability measures P_1 and P_2 on $(\Omega_1, \mathcal{B}_1)$ and $(\Omega_2, \mathcal{B}_2)$, respectively, we try to define on the product space $(\Omega, \mathcal{B})$ a probability measure P by defining for a measurable rectangle $A = A_1 \times A_2$

$$P(A_1 \times A_2) = P_1(A_1) \times P_2(A_2)$$

and extending it to the field $\mathcal{F}$ of finite-disjoint unions of measurable rectangles as the obvious sum.

EXERCISE 1.22. If $E \in \mathcal{F}$ has two representations as finite-disjoint unions of measurable rectangles

$$E = \cup_i (A_1^i \times A_2^i) = \cup_j (B_1^j \times B_2^j),$$

then

$$\sum_i P_1(A_1^i) \times P_2(A_2^i) = \sum_j P_1(B_1^j) \times P_2(B_2^j),$$

so that $P(E)$ is well-defined. P is a finitely additive probability measure on $\mathcal{F}$.

LEMMA 1.10 *The measure P is countably additive on the field $\mathcal{F}$.*

PROOF: For any set $E \in \mathcal{F}$ let us define the section E_{ω_2} as

(1.12) $$E_{\omega_2} = \{\omega_1 : (\omega_1, \omega_2) \in E\}.$$

Then $P_1(E_{\omega_2})$ is a measurable function of ω_2 (is in fact, a simple function) and

(1.13) $$P(E) = \int_{\Omega_2} P_1(E_{\omega_2}) dP_2.$$

Now let $E_n \in \mathcal{F} \downarrow \emptyset$, the empty set. Then it is easy to verify that E_{n,ω_2} defined by

$$E_{n,\omega_2} = \{\omega_1 : (\omega_1, \omega_2) \in E_n\}$$

1.5. PRODUCT SPACES

satisfies $E_{n,\omega_2} \downarrow \emptyset$ for each $\omega_2 \in \Omega_2$. From the countable additivity of P_1 we conclude that $P_1(E_{n,\omega_2}) \to 0$ for each $\omega_2 \in \Omega_2$ and, since $0 \leq P_1(E_{n,\omega_2}) \leq 1$ for $n \geq 1$, it follows from equation (1.13) and the bounded convergence theorem that

$$P(E_n) = \int_{\Omega_2} P_1(E_{n,\omega_2}) dP_2 \to 0$$

establishing the countable additivity of P on $\mathcal{F}$. □

By an application of the Caratheodory extension theorem we conclude that P extends uniquely as a countably additive measure to the σ-field $\mathcal{B}$ (product σ-field) generated by $\mathcal{F}$. We will call this the *product measure P*.

COROLLARY 1.11 *For any* $A \in \mathcal{B}$ *if we denote by* A_{ω_1} *and* A_{ω_2} *the respective sections*

$$A_{\omega_1} = \{\omega_2 : (\omega_1, \omega_2) \in A\} \quad \text{and} \quad A_{\omega_2} = \{\omega_1 : (\omega_1, \omega_2) \in A\},$$

then the functions $P_1(A_{\omega_2})$ *and* $P_2(A_{\omega_1})$ *are* measurable *and*

$$P(A) = \int P_1(A_{\omega_2}) dP_2 = \int P_2(A_{\omega_1}) dP_1 .$$

In particular, for a measurable set A, $P(A) = 0$ *if and only if for almost all* ω_1 *with respect to* P_1*, the sections* A_{ω_1} *have measure* 0 *or equivalently for almost all* ω_2 *with respect to* P_2*, the sections* A_{ω_2} *have measure* 0.

PROOF: The assertion is clearly valid if A is a rectangle of the form $A_1 \times A_2$ with $A_1 \in \mathcal{B}_1$ and $A_2 \in \mathcal{B}_2$. If $A \in \mathcal{F}$, then it is a finite disjoint union of such rectangles and the assertion is extended to such a set by simple addition. Clearly, by the monotone convergence theorem, the class of sets for which the assertion is valid is a monotone class, and since it contains the field $\mathcal{F}$, it also contains the σ-field $\mathcal{B}$ generated by the field $\mathcal{F}$. □

Warning. It is possible that a set A may not be *measurable* with respect to the product σ-field, but nevertheless the sections A_{ω_1} and A_{ω_2} are all measurable, $P_2(A_{\omega_1})$ and $P_1(A_{\omega_2})$ are measurable functions, but

$$\int P_1(A_{\omega_2}) dP_2 \neq \int P_2(A_{\omega_1}) dP_1 .$$

In fact, there is a rather nasty example where $P_1(A_{\omega_2})$ is identically 1 whereas $P_2(A_{\omega_1})$ is identically 0.

The next result concerns the equality of the double integral (i.e., the integral with respect to the product measure) and the repeated integrals in any order.

THEOREM 1.12 (Fubini's Theorem) *Let* $f(\omega) = f(\omega_1, \omega_2)$ *be a measurable function of ω on* $(\Omega, \mathcal{B})$. *Then f can be considered a function of ω_2 for each fixed ω_1 or the other way around. The functions* $g_{\omega_1}(\cdot)$ *and* $h_{\omega_2}(\cdot)$ *defined, respectively, on Ω_2 and Ω_1 by*

$$g_{\omega_1}(\omega_2) = h_{\omega_2}(\omega_1) = f(\omega_1, \omega_2)$$

1. MEASURE THEORY

are measurable for each ω_1 and ω_2. If f is integrable, then the functions $g_{\omega_1}(\omega_2)$ and $h_{\omega_2}(\omega_1)$ are integrable for almost all ω_1 and ω_2, respectively. Their integrals

$$G(\omega_1) = \int_{\Omega_2} g_{\omega_1}(\omega_2) dP_2 \quad \text{and} \quad H(\omega_2) = \int_{\Omega_1} h_{\omega_2}(\omega_1) dP_1$$

are measurable, finite almost everywhere, and integrable with respect to P_1 and P_2, respectively. Finally,

$$\int f(\omega_1, \omega_2) dP = \int G(\omega_1) dP_1 = \int H(\omega_2) dP_2.$$

Conversely, for a nonnegative measurable function f, if either G or H, which are always measurable, has a finite integral, so does the other and f is integrable with its integral being equal to either of the repeated integrals, namely, integrals of G and H.

PROOF: The proof follows the standard pattern. It is a restatement of the earlier corollary if f is the indicator function of a measurable set A. By linearity it is true for simple functions, and by passing to uniform limits it is true for bounded measurable functions f. By monotone limits it is true for nonnegative functions, and, finally, by taking the positive and negative parts separately, it is true for any arbitrary integrable function f. □

Warning. The following could happen: f is a measurable function that takes both positive and negative nonintegrable values. Both the repeated integrals exist and are unequal. The example is not hard.

EXERCISE 1.23. Construct a measurable function $f(x, y)$ which is not integrable on the product $[0, 1] \times [0, 1]$ of two copies of the unit interval with Lebesgue measure such that the repeated integrals make sense and are unequal, i.e.,

$$\int_0^1 dx \int_0^1 f(x, y) dy \neq \int_0^1 dy \int_0^1 f(x, y) dx.$$

1.6. Distributions and Expectations

Let us recall that a triplet $(\Omega, \mathcal{B}, P)$ is a probability space if Ω is a set, $\mathcal{B}$ is a σ-field of subsets of Ω, and P is a (countably additive) probability measure on $\mathcal{B}$. A random variable X is a real-valued measurable function on $(\Omega, \mathcal{B})$. Given such a function X, it induces a probability distribution α on the Borel subsets of the line $\alpha = PX^{-1}$. The distribution function $F(x)$ corresponding to α is obviously

$$F(x) = \alpha((-\infty, x]) = P[\omega : X(\omega) \leq x].$$

The measure α is called the *distribution* of X and $F(x)$ is called the *distribution function* of X. If g is a measurable function of the real variable x, then $Y(\omega) = g(X(\omega))$ is again a random variable and its distribution $\beta = PY^{-1}$ can be obtained

1.6. DISTRIBUTIONS AND EXPECTATIONS

as $\beta = \alpha g^{-1}$ from α. The *expectation* or *mean* of a random variable is defined if it is integrable and

$$E[X] = E^P[X] = \int X(\omega) dP.$$

By the change-of-variables formula (Exercise 3.3) it can be obtained directly from α as

$$E[X] = \int x \, d\alpha.$$

Here we are taking advantage of the fact that on the real line x is a very special real-valued function. The value of the integral in this context is referred to as the expectation or mean of α. Of course, it exists if and only if

$$\int |x| \, d\alpha < \infty \quad \text{and} \quad \left| \int x \, d\alpha \right| \leq \int |x| \, d\alpha.$$

Similarly,

$$E(g(X)) = \int g(X(\omega)) dP = \int g(x) \, d\alpha,$$

and anything concerning X can be calculated from α. The statement that X is a random variable with distribution α has to be interpreted in the sense that somewhere in the background there is a probability space and a random variable X on it which has α for its distribution. Usually only α matters and the underlying $(\Omega, \mathcal{B}, P)$ never emerges from the background, and in a pinch we can always say Ω is the real line, $\mathcal{B}$ are the Borel sets, P is nothing but α, and the random variable $X(x) = x$. Some other related quantities are

(1.14) $$\text{Var}(X) = \sigma^2(X) = E[X^2] - [E[X]]^2.$$

Var(X) is called the *variance* of X.

EXERCISE 1.24. Show that, if it is defined, Var(X) is always nonnegative and Var(X) = 0 if and only if for some value a, which is necessarily equal to $E[X]$, $P[X = a] = 1$.

Somewhat more generally we can consider a measurable mapping $X = (X_1, X_2, \ldots, X_n)$ of a probability space $(\Omega, \mathcal{B}, P)$ into $\mathbb{R}^n$ as a vector of n random variables $X_1(\omega), X_2(\omega), \ldots, X_n(\omega)$. These are called *random vectors* or *vector-valued random variables* and the induced distribution $\alpha = PX^{-1}$ on $\mathbb{R}^n$ is called the *distribution* of X or the *joint distribution* of $(X_1, X_2, \ldots, X_n)$. If we denote by π_i the coordinate maps $(x_1, x_2, \ldots, x_n) \to x_i$ from $\mathbb{R}^n \to \mathbb{R}$, then

$$\alpha_i = \alpha \pi_i^{-1} = PX_i^{-1}$$

are called the *marginals* of α.

The *covariance* between two random variables X and Y is defined as

(1.15) $$\text{Cov}(X, Y) = E[(X - E(X))(Y - E(Y))] = E[XY] - E[X]E[Y].$$

EXERCISE 1.25. If $X_1, X_2, \ldots, X_n$ are n random variables, the matrix
$$C_{i,j} = \text{Cov}(X_i, X_j)$$
is called the *covariance matrix*. Show that it is a symmetric positive semidefinite matrix. Is every positive semidefinite matrix the covariance matrix of some random vector?

EXERCISE 1.26. The Riemann-Stieltjes integral uses the distribution function directly to define $\int_{-\infty}^{\infty} g(x) dF(x)$ where g is a bounded continuous function and F is a distribution function. It is defined as the limit as $N \to \infty$ of sums
$$\sum_{j=0}^{N} g(x_j) [F(a_{j+1}^N) - F(a_j^N)]$$
where $-\infty < a_0^N < a_1^N < \cdots < a_N^N < a_{N+1}^N < \infty$ is a partition of the finite interval $[a_0^N, a_{N+1}^N]$ and the limit is taken in such a way that $a_0^N \to -\infty$, $a_{N+1}^N \to +\infty$, and the oscillation of g in any $[a_j^N, a_{j+1}^N]$ goes to 0. Show that if P is the measure corresponding to F, then
$$\int_{-\infty}^{\infty} g(x) dF(x) = \int_{\mathbb{R}} g(x) dP.$$

CHAPTER 2

Weak Convergence

2.1. Characteristic Functions

If α is a probability distribution on the line, its characteristic function is defined by

(2.1) $$\phi(t) = \int \exp[itx] d\alpha.$$

The above definition makes sense. We write the integrand e^{itx} as $\cos tx + i \sin tx$ and integrate each part to see that

$$|\phi(t)| \leq 1 \quad \text{for all real } t.$$

EXERCISE 2.1. Calculate the characteristic functions for the following distributions:
  (1) α is the degenerate distribution δ_a with probability 1 at the point a.
  (2) α is the binomial distribution with probabilities
  $$p_k = \text{Prob}[X = k] = \binom{n}{k} p^k (1-p)^{n-k} \quad \text{for } 0 \leq k \leq n.$$

THEOREM 2.1 *The characteristic function $\phi(t)$ of any probability distribution is a uniformly continuous function of t that is positive definite, i.e., for any n complex numbers $\xi_1, \xi_2, \ldots, \xi_n$ and real numbers $t_1, t_2, \ldots, t_n$*

$$\sum_{i,j=1}^{n} \phi(t_i - t_j) \xi_i \bar{\xi}_j \geq 0.$$

PROOF: Let us note that

$$\sum_{i,j=1}^{n} \phi(t_i - t_j) \xi_i \bar{\xi}_j = \sum_{i,j=1}^{n} \xi_i \bar{\xi}_j \int \exp[i(t_i - t_j)x] d\alpha$$

$$= \int \left| \sum_{j=1}^{n} \xi_j \exp[it_j x] \right|^2 d\alpha \geq 0.$$

To prove uniform continuity we see that

$$|\phi(t) - \phi(s)| \leq \int |\exp[i(t-s)x] - 1| dP$$

which tends to 0 by the bounded convergence theorem if $|t - s| \to 0$. □

The characteristic function of course carries some information about the distribution α. In particular, if $\int |x|d\alpha < \infty$, then $\phi(\cdot)$ is continuously differentiable and $\phi'(0) = i \int x\, d\alpha$.

EXERCISE 2.2. Prove it!

Warning. The converse need not be true. $\phi(\cdot)$ can be continuously differentiable but $\int |x|dP$ could be ∞.

EXERCISE 2.3. Construct a counterexample along the following lines: Take a discrete distribution that is symmetric around 0 with

$$\alpha\{n\} = \alpha\{-n\} = p(n) \simeq \frac{1}{n^2 \log n}.$$

Then show that $\sum_n \frac{(1-\cos nt)}{n^2 \log n}$ is a continuously differentiable function of t.

EXERCISE 2.4. The story with higher moments $m_r = \int x^r\, d\alpha$ is similar. If any of them, say m_r, exist, then $\phi(\cdot)$ is r times continuously differentiable and $\phi^{(r)}(0) = i^r m_r$. The converse is false for odd r, but true for even r by an application of Fatou's lemma.

The next question is how to recover the distribution function $F(x)$ from $\phi(t)$. If we go back to the Fourier inversion formula (see, for instance, [2]) we can "guess," using the fundamental theorem of calculus and Fubini's theorem, that

$$F'(x) = \frac{1}{2\pi} \int_{-\infty}^{\infty} \exp[-itx]\phi(t)dt$$

and therefore

$$F(b) - F(a) = \frac{1}{2\pi} \int_a^b dx \int_{-\infty}^{\infty} \exp[-itx]\phi(t)dt$$

$$= \frac{1}{2\pi} \int_{-\infty}^{\infty} \phi(t)dt \int_a^b \exp[-itx]dx$$

$$= \frac{1}{2\pi} \int_{-\infty}^{\infty} \phi(t) \frac{\exp[-itb] - \exp[-ita]}{-it} dt$$

$$= \lim_{T \to \infty} \frac{1}{2\pi} \int_{-T}^{T} \phi(t) \frac{\exp[-itb] - \exp[-ita]}{-it} dt.$$

We will in fact prove the final relation, which is a principal value integral, provided a and b are points of continuity of F. We compute the right-hand side as

$$\lim_{T \to \infty} \frac{1}{2\pi} \int_{-T}^{T} \frac{\exp[-itb] - \exp[-ita]}{-it} dt \int \exp[itx]d\alpha$$

$$= \lim_{T \to \infty} \frac{1}{2\pi} \int d\alpha \int_{-T}^{T} \frac{\exp[it(x-b)] - \exp[it(x-a)]}{-it} dt$$

2.1. CHARACTERISTIC FUNCTIONS

$$= \lim_{T \to \infty} \frac{1}{2\pi} \int d\alpha \int_{-T}^{T} \frac{\sin t(x-a) - \sin t(x-b)}{t} dt$$

$$= \frac{1}{2} \int [\text{sign}(x-a) - \text{sign}(x-b)] d\alpha$$

$$= F(b) - F(a)$$

provided a and b are continuity points. We have applied Fubini's theorem and the bounded convergence theorem to take the limit as $T \to \infty$. Note that the Dirichlet integral

$$u(T, z) = \int_0^T \frac{\sin tz}{t} dt$$

satisfies $\sup_{T,z} |u(T,z)| \leq C$ and

$$\lim_{T \to \infty} u(T, z) = \begin{cases} 1 & \text{if } z > 0 \\ -1 & \text{if } z < 0 \\ 0 & \text{if } z = 0. \end{cases}$$

As a consequence we conclude that the distribution function and hence α is determined uniquely by the characteristic function.

EXERCISE 2.5. Prove that if two distribution functions agree on the set of points at which they are both continuous, they agree everywhere.

Besides those in Exercise 2.1, some additional examples of probability distributions and the corresponding characteristic functions are given below:

(1) The Poisson distribution of "rare events" with rate λ has probabilities $P[X = r] = e^{-\lambda} \frac{\lambda^r}{r!}$ for $r \geq 0$. Its characteristic function is

$$\phi(t) = \exp[\lambda(e^{it} - 1)].$$

(2) The geometric distribution, the distribution of the number of unsuccessful attempts preceding a success, has $P[X = r] = pq^r$ for $r \geq 0$. Its characteristic function is

$$\phi(t) = p(1 - qe^{it})^{-1}.$$

(3) The negative binomial distribution, the probability distribution of the number of accumulated failures before k successes with $P[X = r] = \binom{k+r-1}{r} p^k q^r$ has the characteristic function

$$\phi(t) = p^k (1 - qe^{it})^{-k}.$$

We now turn to some common continuous distributions, in fact given by "densities" $f(x)$, i.e., the distribution functions are given by $F(x) = \int_{-\infty}^{x} f(y) dy$.

(1) The "uniform" distribution with density $f(x) = \frac{1}{b-a}, a \leq x \leq b$, has the characteristic function

$$\phi(t) = \frac{e^{itb} - e^{ita}}{it(b-a)}.$$

In particular, for the case of a symmetric interval $[-a, a]$,

$$\phi(t) = \frac{\sin at}{at}.$$

(2) The gamma distribution with density $f(x) = \frac{c^p}{\Gamma(p)} e^{-cx} x^{p-1}$, $x \geq 0$, has the characteristic function

$$\phi(t) = \left(1 - \frac{it}{c}\right)^{-p}$$

where $c > 0$ is any constant. A special case of the gamma distribution is the exponential distribution that corresponds to $c = p = 1$ with density $f(x) = e^{-x}$ for $x \geq 0$. Its characteristic function is given by

$$\phi(t) = [1 - it]^{-1}.$$

(3) The two-sided exponential with density $f(x) = \frac{1}{2} e^{-|x|}$ has the characteristic function

$$\phi(t) = \frac{1}{1 + t^2}.$$

(4) The Cauchy distribution with density $f(x) = \frac{1}{\pi} \frac{1}{1+x^2}$ has the characteristic function

$$\phi(t) = e^{-|t|}.$$

(5) The normal or Gaussian distribution with mean μ and variance σ^2, which has a density of $\frac{1}{\sqrt{2\pi}\sigma} e^{-(x-\mu)^2/2\sigma^2}$, has the characteristic function given by

$$\phi(t) = e^{it\mu - \frac{\sigma^2 t^2}{2}}.$$

In general, if X is a random variable which has distribution α and a characteristic function $\phi(t)$, the distribution β of $aX + b$ can be written as $\beta(A) = \alpha[x : ax + b \in A]$ and its characteristic function $\psi(t)$ can be expressed as $\psi(t) = e^{itb} \phi(at)$. In particular, the characteristic function of $-X$ is $\phi(-t) = \overline{\phi(t)}$. Therefore, the distribution of X is symmetric around $x = 0$ if and only if $\phi(t)$ is real for all t.

2.2. Moment-Generating Functions

If α is a probability distribution on $\mathbb{R}$, for any integer $k \geq 1$ the moment m_k of α is defined as

(2.2) $$m_k = \int x^k \, d\alpha.$$

Or, equivalently, the kth moment of a random variable X is

(2.3) $$m_k = E[X^k].$$

By convention one takes $m_0 = 1$ even if $P[X = 0] > 0$. We should note that if k is odd, in order for m_k to be defined we must have $E[|X|^k] = \int |x|^k \, d\alpha < \infty$. Given a distribution α, either all the moments exist, or they exist only for $0 \leq k \leq k_0$ for some k_0. It could happen that $k_0 = 0$ as is the case with the Cauchy distribution. If

2.2. MOMENT-GENERATING FUNCTIONS

we know all the moments of a distribution α, we know the expectations $\int p(x)d\alpha$ for every polynomial $p(\cdot)$. Since polynomials $p(\cdot)$ can be used to approximate (by the Stone-Weierstrass theorem) any continuous function, one might hope that, from the moments, one can recover the distribution α. This is not as straightforward as one would hope. If we take a bounded continuous function like $\sin x$, we can find a sequence of polynomials $p_n(x)$ that converges to $\sin x$. But to conclude that

$$\int \sin x\, d\alpha = \lim_{n\to\infty} \int p_n(x)\, d\alpha$$

we need to control the contribution to the integral from large values of x, which is the role of the dominated convergence theorem. If we define $p^*(x) = \sup_n |p_n(x)|$, it would be a big help if $\int p^*(x)d\alpha$ were finite. But the degrees of the polynomials p_n have to increase indefinitely with n because $\sin x$ is a transcendental function. Therefore $p^*(\cdot)$ must grow faster than a polynomial at ∞, and the condition $\int p^*(x)d\alpha < \infty$ may not hold.

In general, it is not true that moments determine the distribution. If we look at it through characteristic functions, it is the problem of trying to recover the function $\phi(t)$ from a knowledge of all of its derivatives at $t = 0$. The Taylor series at $t = 0$ may not yield the function. Of course, we have more information in our hands, like positive definiteness, etc. But still it is likely that moments do not in general determine α. In fact, here is how to construct an example.

We need nonnegative numbers $\{a_n\}, \{b_n\} : n \geq 0$ such that

$$\sum_n a_n e^{kn} = \sum_n b_n e^{kn} = m_k \quad \text{for every } k \geq 0.$$

We can then replace them by $\{\frac{a_n}{m_0}\}, \{\frac{b_n}{m_0}\} : n \geq 0$ so that $\sum_k a_k = \sum_k b_k = 1$ and the two probability distributions

$$P[X = e^n] = a_n \quad \text{and} \quad P[X = e^n] = b_n$$

will have all their moments equal. Once we can find $\{c_n\}$ such that

$$\sum_n c_n e^{nz} = 0 \quad \text{for } z = 0, 1, \ldots,$$

we can take $a_n = \max(c_n, 0)$ and $b_n = \max(-c_n, 0)$ and we will have our example. The goal then is to construct $\{c_n\}$ such that $\sum_n c_n z^n = 0$ for $z = 1, e, e^2, \ldots$. Borrowing from ideas in the theory of a complex variable (see the Weierstrass factorization theorem, [1]), we define

$$C(z) = \prod_{n=0}^{\infty} \left(1 - \frac{z}{e^n}\right)$$

and expand $C(z) = \sum c_n z^n$. Since $C(z)$ is an entire function, the coefficients c_n satisfy $\sum_n |c_n| e^{kn} < \infty$ for every k.

There is, in fact, a positive result as well. If α is such that the moments $m_k = \int x^k d\alpha$ do not grow too fast, then α is determined by m_k.

THEOREM 2.2 Let m_k be such that $\sum_k m_{2k} \frac{a^{2k}}{(2k)!} < \infty$ for some $a > 0$. Then there is at most one distribution α such that $\int x^k \, d\alpha = m_k$.

PROOF: We want to determine the characteristic function $\phi(t)$ of α. First we note that if α has moments m_k satisfying our assumption, then

$$\int \cosh(ax) d\alpha = \sum_k \frac{a^{2k}}{(2k)!} m_{2k} < \infty$$

by the monotone convergence theorem. In particular,

$$\psi(u + it) = \int e^{(u+it)x} \, d\alpha$$

is well-defined as an analytic function of $z = u + it$ in the strip $|u| < a$. From the theory of functions of a complex variable we know that the function $\psi(\cdot)$ is uniquely determined in the strip by its derivatives at 0, i.e., $\{m_k\}$. In particular, $\phi(t) = \psi(0 + it)$ is determined as well. □

2.3. Weak Convergence

One of the basic ideas in establishing limit theorems is the notion of *weak convergence* of a sequence of probability distributions on the line $\mathbb{R}$. Since the role of a probability measure is to assign probabilities to sets, we should expect that if two probability measures are to be close, then they should assign for a given set probabilities that are nearly equal. This suggests the definition

$$d(P_1, P_2) = \sup_{A \in \mathcal{B}} |P_1(A) - P_2(A)|$$

as the distance between two probability measures P_1 and P_2 on a measurable space $(\Omega, \mathcal{B})$. This is too strong. If we take P_1 and P_2 to be degenerate distributions with probability 1 concentrated at two points x_1 and x_2 on the line, one can see that, as soon as $x_1 \neq x_2$, $d(P_1, P_2) = 1$ and the above metric is not sensitive to how close the two points x_1 and x_2 are. It only matters that they are unequal. The problem is not because of the supremum. We can take A to be an interval $[a, b]$ that includes x_1 but omits x_2 and $|P_1(A) - P_2(A)| = 1$. On the other hand, if the endpoints of the interval are kept away from x_1 or x_2 the situation is not that bad. This leads to the following definition:

DEFINITION 2.1 A sequence α_n of probability distributions on $\mathbb{R}$ is said to *converge weakly* to a probability distribution α if

$$\lim_{n \to \infty} \alpha_n[I] = \alpha[I] \quad \text{for any interval } I = [a, b]$$

such that the single point sets a and b have probability 0 under α.

One can state this equivalently in terms of the distribution functions $F_n(x)$ and $F(x)$ corresponding to the measures α_n and α, respectively.

2.3. WEAK CONVERGENCE

DEFINITION 2.2 A sequence α_n of probability measures on the real line $\mathbb{R}$ with distribution functions $F_n(x)$ is said to *converge weakly* to a limiting probability measure α with distribution function $F(x)$ (in symbols $\alpha_n \Rightarrow \alpha$ or $F_n \Rightarrow F$) if

$$\lim_{n \to \infty} F_n(x) = F(x) \quad \text{for every } x \text{ that is a continuity point of } F.$$

EXERCISE 2.6. Prove the equivalence of the two definitions.

REMARK 2.1. One says that a sequence X_n of random variables *converges in law* or *in distribution* to X if the distributions α_n of X_n converge weakly to the distribution α of X.

There are equivalent formulations in terms of expectations and characteristic functions.

THEOREM 2.3 (Lévy-Cramér Continuity Theorem) *The following are equivalent:*

(i) $\alpha_n \Rightarrow \alpha$ or $F_n \Rightarrow F$.
(ii) For every *bounded continuous* function $f(x)$ on $\mathbb{R}$

$$\lim_{n \to \infty} \int_{\mathbb{R}} f(x) d\alpha_n = \int_{\mathbb{R}} f(x) d\alpha.$$

(iii) If $\phi_n(t)$ and $\phi(t)$ are, respectively, the characteristic functions of α_n and α for every real t,

$$\lim_{n \to \infty} \phi_n(t) = \phi(t).$$

PROOF: We first prove (i) $\Rightarrow$ (ii). Let $\varepsilon > 0$ be arbitrary. Find continuity points a and b of F such that $a < b$, $F(a) \leq \varepsilon$, and $1 - F(b) \leq \varepsilon$. Since $F_n(a)$ and $F_n(b)$ converge to $F(a)$ and $F(b)$, for n large enough, $F_n(a) \leq 2\varepsilon$ and $1 - F_n(b) \leq 2\varepsilon$. Divide the interval $[a, b]$ into a finite number $N = N_\delta$ of small subintervals $I_j = (a_j, a_{j+1}]$, $1 \leq j \leq N$, with $a = a_1 < a_2 < \cdots < a_{N+1} = b$ such that all the endpoints $\{a_j\}$ are points of continuity of F and the oscillation of the continuous function f in each I_j is less than a preassigned number δ. Since any continuous function f is uniformly continuous in the closed bounded (compact) interval $[a, b]$, this is always possible for any given $\delta > 0$. Let $h(x) = \sum_{j=1}^N \chi_{I_j} f(a_j)$ be the simple function equal to $f(a_j)$ on I_j and 0 outside $\bigcup_j I_j = (a, b]$. We have $|f(x) - h(x)| \leq \delta$ on $(a, b]$. If $f(x)$ is bounded by M, then

$$(2.4) \quad \left| \int f(x) d\alpha_n - \sum_{j=1}^N f(a_j) [F_n(a_{j+1}) - F_n(a_j)] \right| \leq \delta + 4M\varepsilon$$

and

$$(2.5) \quad \left| \int f(x) d\alpha - \sum_{j=1}^N f(a_j) [F(a_{j+1}) - F(a_j)] \right| \leq \delta + 2M\varepsilon.$$

Since $\lim_{n\to\infty} F_n(a_j) = F(a_j)$ for every $1 \le j \le N$, we conclude from equations (2.4) and (2.5) and from the triangle inequality that

$$\limsup_{n\to\infty} \left| \int f(x) d\alpha_n - \int f(x) d\alpha \right| \le 2\delta + 6M\varepsilon.$$

Since ε and δ are arbitrary small numbers we are done. □

Because we can make the choice of $f(x) = \exp[itx] = \cos tx + i\sin tx$, which for every t is a bounded and continuous function, (ii) $\Rightarrow$ (iii) is trivial.

(iii) $\Rightarrow$ (i) is the hardest. It is carried out in several steps. Actually we will prove a stronger version as a separate theorem.

THEOREM 2.4 *For each $n \ge 1$, let $\phi_n(t)$ be the characteristic function of a probability distribution α_n. Assume that $\lim_{n\to\infty} \phi_n(t) = \phi(t)$ exists for each t and $\phi(t)$ is continuous at $t = 0$. Then $\phi(t)$ is the characteristic function of some probability distribution α and $\alpha_n \Rightarrow \alpha$.*

PROOF: The proof has five steps.

Step 1. Let $r_1, r_2, \ldots,$ be an enumeration of the rational numbers. For each j consider the sequence $\{F_n(r_j) : n \ge 1\}$ where F_n is the distribution function corresponding to $\phi_n(\cdot)$. It is a sequence bounded by 1 and we can extract a subsequence that converges. By the diagonalization process we can choose a subsequence $G_k = F_{n_k}$ such that

$$\lim_{k\to\infty} G_k(r) = b_r$$

exists for every rational number r. From the monotonicity of F_n in x we conclude that if $r_1 < r_2$, then $b_{r_1} \le b_{r_2}$.

Step 2. From the skeleton b_r we reconstruct a right continuous monotone function $G(x)$. We define

$$G(x) = \inf_{r > x} b_r.$$

Clearly, if $x_1 < x_2$, then $G(x_1) \le G(x_2)$ and therefore G is nondecreasing. If $x_n \downarrow x$, any $r > x$ satisfies $r > x_n$ for sufficiently large n. This allows us to conclude that $G(x) = \inf_n G(x_n)$ for any sequence $x_n \downarrow x$, proving that $G(x)$ is right continuous.

Step 3. Next we show that at any continuity point x of G

$$\lim_{n\to\infty} G_n(x) = G(x).$$

Let $r > x$ be a rational number. Then $G_n(x) \le G_n(r)$ and $G_n(r) \to b_r$ as $n \to \infty$. Hence,

$$\limsup_{n\to\infty} G_n(x) \le b_r.$$

This is true for every rational $r > x$, and therefore taking the infimum over $r > x$

$$\limsup_{n\to\infty} G_n(x) \le G(x).$$

Suppose now that we have $y < x$. Find a rational r such that $y < r < x$.

$$\liminf_{n\to\infty} G_n(x) \ge \liminf_{n\to\infty} G_n(r) = b_r \ge G(y).$$

2.3. WEAK CONVERGENCE

As this is true for every $y < x$,

$$\liminf_{n\to\infty} G_n(x) \geq \sup_{y<x} G(y) = G(x-0) = G(x),$$

the last step being a consequence of the assumption that x is a point of continuity of G, i.e., $G(x-0) = G(x)$.

Warning. This does not mean that G is necessarily a distribution function. Consider $F_n(x) = 0$ for $x < n$ and 1 for $x \geq n$, which corresponds to the distribution with the entire probability concentrated at n. In this case $\lim_{n\to\infty} F_n(x) = G(x)$ exists and $G(x) \equiv 0$, which is not a distribution function.

Step 4. We will use the continuity at $t = 0$ of $\phi(t)$ to show that G is indeed a distribution function. If $\phi(t)$ is the characteristic function of α

$$\frac{1}{2T}\int_{-T}^{T} \phi(t)dt = \int \left[\frac{1}{2T}\int_{-T}^{T} \exp[itx]dt\right]d\alpha$$

$$= \int \frac{\sin Tx}{Tx} d\alpha$$

$$\leq \int \left|\frac{\sin Tx}{Tx}\right| d\alpha$$

$$= \int_{|x|<\ell} \left|\frac{\sin Tx}{Tx}\right| d\alpha + \int_{|x|\geq \ell} \left|\frac{\sin Tx}{Tx}\right| d\alpha$$

$$\leq \alpha[|x| < \ell] + \frac{1}{T\ell}\alpha[|x| \geq \ell].$$

We have used Fubini's theorem in the first line and the bounds $|\sin x| \leq |x|$ and $|\sin x| \leq 1$ in the last line. We can rewrite this as

$$1 - \frac{1}{2T}\int_{-T}^{T}\phi(t)dt \geq 1 - \alpha[|x| < \ell] - \frac{1}{T\ell}\alpha[|x| \geq \ell]$$

$$= \alpha[|x| \geq \ell] - \frac{1}{T\ell}\alpha[|x| \geq \ell]$$

$$= \left(1 - \frac{1}{T\ell}\right)\alpha[|x| \geq \ell]$$

$$\geq \left(1 - \frac{1}{T\ell}\right)[1 - F(\ell) + F(-\ell)].$$

Finally, if we pick $\ell = \frac{2}{T}$,

$$\left[1 - F\left(\frac{2}{T}\right) + F\left(-\frac{2}{T}\right)\right] \leq 2\left[1 - \frac{1}{2T}\int_{-T}^{T}\phi(t)dt\right].$$

Since this inequality is valid for any distribution function F and its characteristic function ϕ, we conclude that, for every $k \geq 1$,

(2.6) $$\left[1 - F_{n_k}\left(\frac{2}{T}\right) + F_{n_k}\left(-\frac{2}{T}\right)\right] \leq 2\left[1 - \frac{1}{2T}\int_{-T}^{T}\phi_{n_k}(t)dt\right].$$

We can pick T such that $\pm\frac{2}{T}$ are continuity points of G. If we now pass to the limit and use the bounded convergence theorem on the right-hand side of equation (2.6), we obtain

$$\left[1 - G\left(\frac{2}{T}\right) + G\left(-\frac{2}{T}\right)\right] \leq 2\left[1 - \frac{1}{2T}\int_{-T}^{T}\phi(t)dt\right].$$

Since $\phi(0) = 1$ and ϕ is continuous at $t = 0$, by letting $T \to 0$ in such a way that $\pm\frac{2}{T}$ are continuity points of G, we conclude that

$$1 - G(\infty) + G(-\infty) = 0$$

or G is indeed a distribution function.

Step 5. We now complete the rest of the proof, i.e., show that $\alpha_n \Rightarrow \alpha$. We have $G_k = F_{n_k} \Rightarrow G$ as well as $\psi_k = \phi_{n_k} \to \phi$. Therefore G must equal F, which has ϕ for its characteristic function. Since the argument works for any subsequence of F_n, every subsequence of F_n will have a further subsequence that converges weakly to the same limit F uniquely determined as the distribution function whose characteristic function is $\phi(\cdot)$. Consequently $F_n \Rightarrow F$ or $\alpha_n \Rightarrow \alpha$. □

EXERCISE 2.7. How do you actually prove that if every subsequence of a sequence $\{F_n\}$ has a further subsequence that converges to a common F, then $F_n \Rightarrow F$?

DEFINITION 2.3 A subset $\mathcal{A}$ of probability distributions on $\mathbb{R}$ is said to be *totally bounded* if, given any sequence α_n from $\mathcal{A}$, there is a subsequence that converges weakly to some limiting probability distribution α.

THEOREM 2.5 *In order that a family $\mathcal{A}$ of probability distributions be totally bounded, it is necessary and sufficient that either of the following equivalent conditions hold:*

(2.7) $$\lim_{\ell \to \infty} \sup_{\alpha \in \mathcal{A}} \alpha[x : |x| \geq \ell] = 0,$$

(2.8) $$\lim_{h \to 0} \sup_{\alpha \in \mathcal{A}} \sup_{|t| \leq h} |1 - \phi_\alpha(t)| = 0.$$

Here $\phi_\alpha(t)$ is the characteristic function of α.
The condition in equation (2.7) is often called the *uniform tightness property*.

PROOF: The proof is already contained in the details of the proof of the earlier theorem. We can always choose a subsequence such that the distribution functions converge at rationals and try to reconstruct the limiting distribution function from the limits at rationals. The crucial step is to prove that the limit is a distribution function. Either of the two conditions (2.7) or (2.8) will guarantee this. If condition (2.7) is violated, it is straightforward to pick a sequence from $\mathcal{A}$ for which the distribution functions have a limit which is not a distribution function. Then $\mathcal{A}$ cannot be totally bounded. Condition (2.7) is therefore necessary. That (2.7) ⇒ (2.8) is a

2.3. WEAK CONVERGENCE

consequence of the estimate

$$|1 - \phi(t)| \leq \int |\exp[itx] - 1| d\alpha$$

$$= \int_{|x| \leq \ell} |\exp[itx] - 1| d\alpha + \int_{|x| > \ell} |\exp[itx] - 1| d\alpha$$

$$\leq |t|\ell + 2\alpha[x : |x| > \ell].$$

It is a well-known principle in Fourier analysis that the regularity of $\phi(t)$ at $t = 0$ is related to the decay rate of the tail probabilities.

EXERCISE 2.8. Compute $\int |x|^p d\alpha$ in terms of the characteristic function $\phi(t)$ for p in the range $0 < p < 2$.

Hint. Look at the formula

$$\int_{-\infty}^{\infty} \frac{1 - \cos tx}{|t|^{p+1}} dt = C_p |x|^p$$

and use Fubini's theorem.

We have the following result on the behavior of $\alpha_n(A)$ for certain sets whenever $\alpha_n \Rightarrow \alpha$:

THEOREM 2.6 *Let* $\alpha_n \Rightarrow \alpha$ *on* $\mathbb{R}$. *If* $C \subset \mathbb{R}$ *is a closed set, then*

$$\limsup_{n \to \infty} \alpha_n(C) \leq \alpha(C),$$

while for open sets $G \subset \mathbb{R}$

$$\liminf_{n \to \infty} \alpha_n(G) \geq \alpha(G).$$

If $A \subset \mathbb{R}$ *is a continuity set of* α, *i.e.,* $\alpha(\partial A) = \alpha(\bar{A} - A^o) = 0$, *then*

$$\lim_{n \to \infty} \alpha_n(A) = \alpha(A).$$

PROOF: The function $d(x, C) = \inf_{y \in C} |x - y|$ is continuous and equals 0 precisely on C.

$$f(x) = \frac{1}{1 + d(x, C)}$$

is a continuous function bounded by 1 that is equal to 1 precisely on C, and

$$f_k(x) = [f(x)]^k \downarrow \chi_C(x) \quad \text{as } k \to \infty.$$

For every $k \geq 1$, we have

$$\lim_{n \to \infty} \int f_k(x) d\alpha_n = \int f_k(x) d\alpha$$

and therefore

$$\limsup_{n \to \infty} \alpha_n(C) \leq \lim_{n \to \infty} \int f_k(x) d\alpha_n = \int f_k(x) d\alpha.$$

Letting $k \to \infty$ we get

$$\limsup_{n \to \infty} \alpha_n(C) \leq \alpha(C).$$

Taking complements we conclude that for any open set $G \subset \mathbb{R}$

$$\liminf_{n \to \infty} \alpha_n(G) \geq \alpha(G).$$

Combining the two parts, if $A \subset \mathbb{R}$ is a continuity set of α, i.e., $\alpha(\partial A) = \alpha(\bar{A} - A^o) = 0$, then

$$\lim_{n \to \infty} \alpha_n(A) = \alpha(A).$$

□

We are now ready to prove the converse of Theorem 2.1, which is the hard part of a theorem of Bochner that characterizes the characteristic functions of probability distributions as continuous positive definite functions on $\mathbb{R}$ normalized to be 1 at 0.

THEOREM 2.7 (Bochner's Theorem) *If $\phi(t)$ is a positive definite function which is continuous at $t = 0$ and is normalized so that $\phi(0) = 1$, then ϕ is the characteristic function of some probability distribution on $\mathbb{R}$.*

PROOF: The proof depends on constructing approximations $\phi_n(t)$ which are in fact characteristic functions and satisfy $\phi_n(t) \to \phi(t)$ as $n \to \infty$. Then we can apply the preceding theorem, and the probability measures corresponding to ϕ_n will have a weak limit which will have ϕ for its characteristic function. The proof has three steps.

Step 1. Let us establish a few elementary properties of positive definite functions.

(1) If $\phi(t)$ is a positive definite function, so is $\phi(t) \exp[ita]$ for any real a. The proof is elementary and requires just direct verification.
(2) If $\phi_j(t)$ are positive definite for each j, then so is any linear combination $\phi(t) = \sum_j w_j \phi_j(t)$ with nonnegative weights w_j. If each $\phi_j(t)$ is normalized with $\phi_j(0) = 1$ and $\sum_j w_j = 1$, then of course $\phi(0) = 1$ as well.
(3) If ϕ is positive definite, then ϕ satisfies $\phi(-t) = \overline{\phi(t)}$ and $|\phi(t)| \leq 1 = \phi(0)$ for all t.

We use the fact that the matrix $\{\phi(t_i - t_j) : 1 \leq i, j \leq n\}$ is Hermitian positive definite for any n real numbers $t_1, t_2, \ldots, t_n$. The first assertion follows from the positivity of $\phi(0)|z|^2$; the second is a consequence of the Hermitian property; and if we take $n = 2$ with $t_1 = t$ and $t_2 = 0$, as a consequence of the positive definiteness of the 2×2 matrix we get $|\phi(t)|^2 \leq 1$.

(4) For any s, t we have $|\phi(t) - \phi(s)|^2 \leq 4\phi(0)|1 - \phi(t-s)|$.

We use the positive definiteness of the 3×3 matrix

$$\begin{bmatrix} 1 & \phi(t-s) & \phi(t) \\ \overline{\phi(t-s)} & 1 & \phi(s) \\ \overline{\phi(t)} & \overline{\phi(s)} & 1 \end{bmatrix}$$

2.3. WEAK CONVERGENCE

which is $\{\phi(t_i - t_j)\}$ with $t_1 = t$, $t_2 = s$, and $t_3 = 0$. In particular, the determinant has to be nonnegative.

$$0 \leq 1 + \phi(s)\overline{\phi(t-s)\phi(t)} + \overline{\phi(s)}\phi(t-s)\phi(t) - |\phi(s)|^2 - |\phi(t)|^2 - |\phi(t-s)|^2$$
$$= 1 - |\phi(s) - \phi(t)|^2 - |\phi(t-s)|^2 - \phi(t)\overline{\phi(s)}(1 - \phi(t-s))$$
$$- \overline{\phi(t)}\phi(s)(1 - \overline{\phi(t-s)})$$
$$\leq 1 - |\phi(s) - \phi(t)|^2 - |\phi(t-s)|^2 + 2|1 - \phi(t-s)|,$$

or

$$|\phi(s) - \phi(t)|^2 \leq 1 - |\phi(s-t)|^2 + 2|1 - \phi(t-s)| \leq 4|1 - \phi(s-t)|.$$

(5) It now follows from (4) that if a positive definite function is continuous at $t = 0$, it is continuous everywhere (in fact, uniformly continuous).

Step 2. First we show that if $\phi(t)$ is a positive definite function which is continuous on $\mathbb{R}$ and is absolutely integrable, then

$$f(x) = \frac{1}{2\pi} \int_{-\infty}^{\infty} \exp[-itx]\phi(t)dt \geq 0$$

is a continuous function and

$$\int_{-\infty}^{\infty} f(x)dx = 1.$$

Moreover, the function

$$F(x) = \int_{-\infty}^{x} f(y)dy$$

defines a distribution function with characteristic function

(2.9) $$\phi(t) = \int_{-\infty}^{\infty} \exp[itx]f(x)dx.$$

If ϕ is integrable on $(-\infty, \infty)$, then $f(x)$ is clearly bounded and continuous. To see that it is nonnegative we write

(2.10) $$f(x) = \lim_{T \to \infty} \frac{1}{2\pi} \int_{-T}^{T} \left(1 - \frac{|t|}{T}\right) e^{-itx} \phi(t)dt$$

(2.11) $$= \lim_{T \to \infty} \frac{1}{2\pi T} \int_{0}^{T} \int_{0}^{T} e^{-i(t-s)x} \phi(t-s)dt\,ds$$

(2.12) $$= \lim_{T \to \infty} \frac{1}{2\pi T} \int_{0}^{T} \int_{0}^{T} e^{-itx} e^{isx} \phi(t-s)dt\,ds \geq 0.$$

We can use the dominated convergence theorem to prove equation (2.10), a change of variables to show equation (2.11), and, finally, a Riemann sum approximation to the integral and the positive definiteness of ϕ to show that the quantity in (2.12) is nonnegative. It remains to show the relation (2.9). Let us define

$$f_\sigma(x) = f(x) \exp\left[-\frac{\sigma^2 x^2}{2}\right]$$

and calculate for $t \in \mathbb{R}$, using Fubini's theorem

$$\int_{-\infty}^{\infty} e^{itx} f_\sigma(x) dx = \int_{-\infty}^{\infty} e^{itx} f(x) \exp\left[-\frac{\sigma^2 x^2}{2}\right] dx$$

$$= \frac{1}{2\pi} \int_{-\infty}^{\infty} \int_{-\infty}^{\infty} e^{itx} \phi(s) e^{-isx} \exp\left[-\frac{\sigma^2 x^2}{2}\right] ds\, dx$$

(2.13)
$$= \int_{-\infty}^{\infty} \phi(s) \frac{1}{\sqrt{2\pi}\sigma} \exp\left[-\frac{(t-s)^2}{2\sigma^2}\right] ds.$$

If we take $t = 0$ in equation (2.13), we get

(2.14) $$\int_{-\infty}^{\infty} f_\sigma(x) dx = \int_{-\infty}^{\infty} \phi(s) \frac{1}{\sqrt{2\pi}\sigma} \exp\left[-\frac{s^2}{2\sigma^2}\right] ds \leq 1.$$

Now we let $\sigma \to 0$. Since $f_\sigma \geq 0$ and tends to f as $\sigma \to 0$, from Fatou's lemma and equation (2.14) it follows that f is integrable and in fact, $\int_{-\infty}^{\infty} f(x) dx \leq 1$. Now we let $\sigma \to 0$ in equation (2.13). Since $f_\sigma(x) e^{itx}$ is dominated by the integrable function f, there is no problem with the left-hand side. On the other hand, the limit as $\sigma \to 0$ is easily calculated on the right-hand side of equation (2.13),

$$\int_{-\infty}^{\infty} e^{itx} f(x) dx = \lim_{\sigma \to 0} \int_{-\infty}^{\infty} \phi(s) \frac{1}{\sqrt{2\pi}\sigma} \exp\left[-\frac{(s-t)^2}{2\sigma^2}\right] ds$$

$$= \lim_{\sigma \to 0} \int_{-\infty}^{\infty} \phi(t + \sigma s) \frac{1}{\sqrt{2\pi}} \exp\left[-\frac{s^2}{2}\right] ds = \phi(t),$$

proving equation (2.9).

Step 3. If $\phi(t)$ is a positive definite function which is continuous, so is $\phi(t) \exp[ity]$ for every y and for $\sigma > 0$, as well as the convex combination

$$\phi_\sigma(t) = \int_{-\infty}^{\infty} \phi(t) \exp[ity] \frac{1}{\sqrt{2\pi}\sigma} \exp\left[-\frac{y^2}{2\sigma^2}\right] dy = \phi(t) \exp\left[-\frac{\sigma^2 t^2}{2}\right].$$

The previous step is applicable to $\phi_\sigma(t)$, which is clearly integrable on $\mathbb{R}$, and by letting $\sigma \to 0$ we conclude by Theorem 2.3 that ϕ is a characteristic function as well. □

REMARK 2.2. There is a Fourier series analogue involving distributions on a finite interval, say $S = [0, 2\pi)$. The right endpoint is omitted on purpose, because the distribution should be thought of as being on $[0, 2\pi]$ with 0 and 2π identified. If α is a distribution on S, the characteristic function is defined as

$$\phi(n) = \int e^{inx} d\alpha$$

for integral values $n \in \mathbb{Z}$. There is a uniqueness theorem and a Bochner-type theorem involving an analogous definition of positive definiteness. The proof is nearly the same.

2.3. WEAK CONVERGENCE

EXERCISE 2.9. If $\alpha_n \Rightarrow \alpha$, it is not always true that $\int x \, d\alpha_n \to \int x \, d\alpha$ because, while x is a continuous function, it is not bounded. Construct a simple counterexample. On the positive side, let $f(x)$ be a continuous function that is not necessarily bounded. Assume that there exists a positive continuous function $g(x)$ satisfying

$$\lim_{|x| \to \infty} \frac{|f(x)|}{g(x)} = 0 \quad \text{and} \quad \sup_n \int g(x) d\alpha_n \leq C < \infty.$$

Then show that

$$\lim_{n \to \infty} \int f(x) d\alpha_n = \int f(x) d\alpha.$$

In particular, if $\int |x|^k \, d\alpha_n$ remains bounded, then $\int x^j \, d\alpha_n \to \int x^j \, d\alpha$ for $1 \leq j \leq k-1$.

EXERCISE 2.10. On the other hand, if $\alpha_n \Rightarrow \alpha$ and $g : \mathbb{R} \to \mathbb{R}$ is a continuous function, then the distribution β_n of g under α_n defined as

$$\beta_n[A] = \alpha_n[x : g(x) \in A]$$

converges weakly to β, the corresponding distribution of g under α.

EXERCISE 2.11. If $g_n(x)$ is a sequence of continuous functions such that

$$\sup_{n,x} |g_n(x)| \leq C < \infty \quad \text{and} \quad \lim_{n \to \infty} g_n(x) = g(x)$$

uniformly on every bounded interval, then whenever $\alpha_n \Rightarrow \alpha$ it follows that

$$\lim_{n \to \infty} \int g_n(x) d\alpha_n = \int g(x) d\alpha.$$

Can you construct an example to show that even if g_n, g are continuous just the pointwise convergence $\lim_{n \to \infty} g_n(x) = g(x)$ is not enough?

EXERCISE 2.12. If a sequence $\{f_n(\omega)\}$ of random variables on a measure space are such that $f_n \to f$ in measure, then show that the sequence of distributions α_n of f_n on $\mathbb{R}$ converges weakly to the distribution α of f. Give an example to show that the converse is not true in general. However, if f is equal to a constant c with probability 1, or equivalently α is degenerate at some point c, then $\alpha_n \Rightarrow \alpha = \delta_c$ implies the convergence in probability of f_n to the constant function c.

CHAPTER 3

Independent Sums

3.1. Independence and Convolution

One of the central ideas in probability is the notion of independence. In intuitive terms two events are independent if they have no influence on each other. The formal definition is

DEFINITION 3.1 Two events A and B are said to be *independent* if
$$P[A \cap B] = P[A]P[B].$$

EXERCISE 3.1. If A and B are independent prove that so are A^c and B.

DEFINITION 3.2 Two random variables X and Y are *independent* if the events $X \in A$ and $Y \in B$ are independent for any two Borel sets A and B on the line, i.e.,
$$P[X \in A, Y \in B] = P[X \in A]P[Y \in B] \quad \text{for all Borel sets } A \text{ and } B.$$

There is a natural extension to a finite or even an infinite collection of random variables.

DEFINITION 3.3 A finite collection $\{X_j : 1 \leq j \leq n\}$ of random variables is said to be *independent* if for any n Borel sets $A_1, A_2, \ldots, A_n$ on the line
$$P\left[\bigcap_{1 \leq j \leq n} [X_j \in A_j]\right] = \prod_{1 \leq j \leq n} P[X_j \in A_j].$$

DEFINITION 3.4 An infinite collection of random variables is said to be *independent* if every finite subcollection is independent.

LEMMA 3.1 *Two random variables X, Y defined on (Ω, Σ, P) are independent if and only if the measure induced on $\mathbb{R}^2$ by (X, Y) is the product measure $\alpha \times \beta$ where α and β are the distributions on $\mathbb{R}$ induced by X and Y, respectively.*

PROOF: Left as an exercise.

The important thing to note is that if X and Y are independent and one knows their distributions α and β, then their joint distribution is automatically determined as the product measure.

If X and Y are independent random variables having α and β for their distributions, the distribution of the sum $Z = X + Y$ is determined as follows: First we construct the product measure $\alpha \times \beta$ on $\mathbb{R} \times \mathbb{R}$ and then consider the induced distribution of the function $f(x, y) = x + y$. This distribution, called the *convolution* of α and β, is denoted by $\alpha * \beta$. An elementary calculation using Fubini's theorem provides the following identities:

(3.1) $$(\alpha * \beta)(A) = \int \alpha(A - x) d\beta = \int \beta(A - x) d\alpha.$$

In terms of characteristic functions, we can express the characteristic function of the convolution as

$$\int \exp[itx] d(\alpha * \beta) = \iint \exp[it(x + y)] d\alpha \, d\beta$$
$$= \int \exp[itx] d\alpha \int \exp[itx] d\beta,$$

or equivalently,

(3.2) $$\phi_{\alpha*\beta}(t) = \phi_\alpha(t) \phi_\beta(t),$$

which provides a direct way of calculating the distributions of sums of independent random variables by the use of characteristic functions.

EXERCISE 3.2. If X and Y are independent, show that for any two measurable functions f and g, $f(X)$ and $g(Y)$ are independent.

EXERCISE 3.3. Use Fubini's theorem to show that if X and Y are independent and if f and g are measurable functions with both $E[|f(X)|]$ and $E[|g(Y)|]$ finite, then
$$E[f(X)g(Y)] = E[f(X)]E[g(Y)].$$

EXERCISE 3.4. Show that if X and Y are any two random variables, then $E(X + Y) = E(X) + E(Y)$. If X and Y are two independent random variables, then show that
$$\mathrm{Var}(X + Y) = \mathrm{Var}(X) + \mathrm{Var}(Y)$$
where
$$\mathrm{Var}(X) = E\big[[X - E[X]]^2\big] = E[X^2] - [E[X]]^2.$$

If $X_1, X_2, \ldots, X_n$ are n independent random variables, then the distribution of their sum $S_n = X_1 + X_2 + \cdots + X_n$ can be computed in terms of the distributions of the summands. If α_j is the distribution of X_j, then the distribution of μ_n of S_n is given by the convolution $\mu_n = \alpha_1 * \alpha_2 * \cdots * \alpha_n$, which can be calculated inductively by $\mu_{j+1} = \mu_j * \alpha_{j+1}$. In terms of their characteristic functions $\psi_n(t) = \phi_1(t)\phi_2(t)\cdots\phi_n(t)$. The first two moments of S_n are computed easily.
$$E(S_n) = E(X_1) + E(X_2) + \cdots + E(X_n)$$
and
$$\mathrm{Var}(S_n) = E[S_n - E(S_n)]^2$$
$$= \sum_j E[X_j - E(X_j)]^2 + 2 \sum_{1 \le i < j \le n} E[X_i - E(X_i)][X_j - E(X_j)].$$

For $i \neq j$, because of independence
$$E[X_i - E(X_i)][X_j - E(X_j)] = E[X_i - E(X_i)]E[X_j - E(X_j)] = 0,$$

and we get the formula

(3.3) $$\text{Var}(S_n) = \text{Var}(X_1) + \text{Var}(X_2) + \cdots + \text{Var}(X_n).$$

3.2. Weak Law of Large Numbers

Let us look at the distribution of the number of successes in n independent trials, with the probability of success in a single trial being equal to p:

$$P\{S_n = r\} = \binom{n}{r} p^r (1-p)^{n-r}$$

and

$$P\{|S_n - np| \geq n\delta\} = \sum_{|r-np| \geq n\delta} \binom{n}{r} p^r (1-p)^{n-r}$$

(3.4) $$\leq \frac{1}{n^2 \delta^2} \sum_{|r-np| \geq n\delta} (r - np)^2 \binom{n}{r} p^r (1-p)^{n-r}$$

$$\leq \frac{1}{n^2 \delta^2} \sum_{1 \leq r \leq n} (r - np)^2 \binom{n}{r} p^r (1-p)^{n-r}$$

(3.5) $$= \frac{1}{n^2 \delta^2} E[S_n - np]^2 = \frac{1}{n^2 \delta^2} \text{Var}(S_n)$$

(3.6) $$= \frac{1}{n^2 \delta^2} np(1-p)$$

$$= \frac{p(1-p)}{n\delta^2}.$$

In step (3.4) we have used a discrete version of the simple inequality

$$\int_{x:g(x) \geq a} g(x) d\alpha \geq a\alpha[x : g(x) \geq a]$$

with $g(x) = (x - np)^2$ and in (3.6) we have used the fact that $S_n = X_1 + X_2 + \cdots + X_n$ where the X_i are independent and have the simple distribution $P\{X_i = 1\} = p$ and $P\{X_i = 0\} = 1 - p$. Therefore $E(S_n) = np$ and $\text{Var}(S_n) = n\,\text{Var}(X_1) = np(1-p)$.

It follows now that

$$\lim_{n \to \infty} P\{|S_n - np| \geq n\delta\} = \lim_{n \to \infty} P\left\{\left|\frac{S_n}{n} - p\right| \geq \delta\right\} = 0,$$

or the average $\frac{S_n}{n}$ converges to p in probability. This is easily seen to be equivalent to the statement that the distribution of $\frac{S_n}{n}$ converges to the distribution degenerate at p. See (2.12).

The above argument works for any sequence of independent and identically distributed random variables. If we assume that $E(X_i) = m$ and $\text{Var}(X_i) =$

$\sigma^2 < \infty$, then $E(\frac{S_n}{n}) = m$ and $\text{Var}(\frac{S_n}{n}) = \frac{\sigma^2}{n}$. Chebyshev's inequality states that for any random variable X

$$P\{|X - E[X]| \geq \delta\} = \int_{|X-E[X]|\geq\delta} dP \leq \frac{1}{\delta^2}\int_{|X-E[X]|\geq\delta} [X - E[X]]^2 dP$$

$$\leq \frac{1}{\delta^2}\int [X - E[X]]^2 dP$$

(3.7)
$$= \frac{1}{\delta^2}\text{Var}(X).$$

This can be used to prove the weak law of large numbers for the general case of independent identically distributed random variables with finite second moments.

THEOREM 3.2 *If $X_1, X_2, \ldots, X_n, \ldots$ is a sequence of independent identically distributed random variables with $E[X_j] \equiv m$ and $\text{Var}(X_j) \equiv \sigma^2$, then for*

$$S_n = X_1 + X_2 + \cdots + X_n \quad \text{we have} \quad \lim_{n\to\infty} P\left[\left|\frac{S_n}{n} - m\right| \geq \delta\right] = 0 \quad \text{for any } \delta > 0.$$

PROOF: Use Chebyshev's inequality to estimate

$$P\left[\left|\frac{S_n}{n} - m\right| \geq \delta\right] \leq \frac{1}{\delta^2}\text{Var}\left(\frac{S_n}{n}\right) = \frac{\sigma^2}{n\delta^2}.$$

□

Actually, it is enough to assume that $E|X_i| < \infty$ and the existence of the second moment is not needed. We will provide two proofs of the statement.

THEOREM 3.3 *If $X_1, X_2, \ldots, X_n$ are independent and identically distributed with a finite first moment and $E(X_i) = m$, then $\frac{X_1+X_2+\cdots+X_n}{n}$ converges to m in probability as $n \to \infty$.*

FIRST PROOF: Let C be a large constant and let us define X_i^C as the *truncated* random variable $X_i^C = X_i$ if $|X_i| \leq C$ and $X_i^C = 0$ otherwise. Let $Y_i^C = X_i - X_i^C$ so that $X_i = X_i^C + Y_i^C$. Then

$$\frac{1}{n}\sum_{1\leq i\leq n} X_i = \frac{1}{n}\sum_{1\leq i\leq n} X_i^C + \frac{1}{n}\sum_{1\leq i\leq n} Y_i^C = \xi_n^C + \eta_n^C.$$

If we let $a_C = E(X_i^C)$ and $b_C = E(Y_i^C)$ we always have $m = a_C + b_C$. Consider the quantity

(3.8)
$$\delta_n = E\left[\left|\frac{1}{n}\sum_{1\leq i\leq n} X_i - m\right|\right] = E[|\xi_n^C + \eta_n^C - m|]$$
$$\leq E[|\xi_n^C - a_C|] + E[|\eta_n^C - b_C|]$$
$$\leq \left[E[|\xi_n^C - a_C|^2]\right]^{\frac{1}{2}} + 2E[|Y_i^C|].$$

3.2. WEAK LAW OF LARGE NUMBERS

The truncated random variables X_i^C are bounded by C and are mutually independent. Theorem 3.2 is applicable to them, and hence the first of the two terms in (3.8) tends to 0. Taking the lim sup as $n \to \infty$, for any $0 < C < \infty$,

$$\limsup_{n \to \infty} \delta_n \leq 2E[|Y_i^C|].$$

If we now let the cutoff level C go to infinity, by the integrability of X_i, $E[|Y_i^C|] \to 0$ as $C \to \infty$ and we are done. The final step of establishing that, for any sequence Y_n of random variables, $E[|Y_n|] \to 0$ implies that $Y_n \to 0$ in probability, is left as an exercise and is not very different from Chebyshev's inequality. $\square$

SECOND PROOF: We can use characteristic functions. If we denote the characteristic function of X_i by $\phi(t)$, then the characteristic function of $\frac{1}{n}\sum_{1 \leq i \leq n} X_i$ is given by $\psi_n(t) = [\phi(\frac{t}{n})]^n$. The existence of the first moment assures us that $\phi(t)$ is differentiable at $t = 0$ with a derivative equal to im where $m = E(X_i)$. Therefore, by Taylor expansion,

$$\phi\left(\frac{t}{n}\right) = 1 + \frac{imt}{n} + o\left(\frac{1}{n}\right).$$

Whenever $na_n \to z$, it follows that $(1 + a_n)^n \to e^z$. Therefore,

$$\lim_{n \to \infty} \psi_n(t) = \exp[imt],$$

which is the characteristic function of the distribution degenerate at m. Hence, the distribution of $\frac{S_n}{n}$ tends to the degenerate distribution at the point m. The weak law of large numbers is thereby established.

EXERCISE 3.5. If the underlying distribution is a Cauchy distribution with density $\frac{1}{\pi(1+x^2)}$ and characteristic function $\phi(t) = e^{-|t|}$, prove that the weak law does not hold.

EXERCISE 3.6. The weak law may hold sometimes, even if the mean does not exist. If we dampen the tails of the Cauchy ever so slightly with a density $f(x) = \frac{c}{(1+x^2)\log(1+x^2)}$, show that the weak law of large numbers holds.

EXERCISE 3.7. In the case of the binomial distribution with $p = \frac{1}{2}$, use Stirling's formula

$$n! \simeq \sqrt{2\pi} e^{-n} n^{n+\frac{1}{2}}$$

to estimate the probability

$$\sum_{r \geq nx} \binom{n}{r} \frac{1}{2^n}$$

and show that it decays geometrically in n. Can you calculate the geometric ratio

$$\rho(x) = \lim_{n \to \infty} \left[\sum_{r \geq nx} \binom{n}{r} \frac{1}{2^n}\right]^{\frac{1}{n}}$$

explicitly as a function of x for $x > \frac{1}{2}$?

3.3. Strong Limit Theorems

The weak law of large numbers is really a result concerning the behavior of

$$\left(\frac{S_n}{n}\right) = \frac{X_1 + X_2 + \cdots + X_n}{n}$$

where $X_1, X_2, \ldots, X_n, \ldots$ is a sequence of independent and identically distributed random variables on some probability space $(\Omega, \mathcal{B}, P)$. Under the assumption that X_i are integrable with an integral equal to m, the weak law asserts that as $n \to \infty$, $\frac{S_n}{n} \to m$ in probability. Since almost everywhere convergence is generally stronger than convergence in probability, one may ask if

$$P\left[\omega : \lim_{n \to \infty} \frac{S_n(\omega)}{n} = m\right] = 1.$$

This is called the *strong law of large numbers*. Strong laws are statements that hold for almost all ω.

Let us look at functions of the form $f_n = \chi_{A_n}$. It is easy to verify that $f_n \to 0$ in probability if and only if $P(A_n) \to 0$. On the other hand:

LEMMA 3.4 (Borel-Cantelli Lemma) *If*

$$\sum_n P(A_n) < \infty \quad \text{then} \quad P\left[\omega : \lim_{n \to \infty} \chi_{A_n}(\omega) = 0\right] = 1.$$

If the events A_n are mutually independent, the converse is also true.

REMARK 3.1. Note that the complementary event

$$\left[\omega : \limsup_{n \to \infty} \chi_{A_n}(\omega) = 1\right]$$

is the same as $\bigcap_{n=1}^{\infty} \bigcup_{j=n}^{\infty} A_j$, or the event that infinitely many of the events $\{A_j\}$ occur.

The conclusion of the next exercise will be used in the proof.

EXERCISE 3.8. Prove the following variant of the monotone convergence theorem: If $f_n(\omega) \geq 0$ are measurable functions, the set $E = \{\omega : S(\omega) = \sum_n f_n(\omega) < \infty\}$ is measurable and $S(\omega)$ is a measurable function on E. If each f_n is integrable and $\sum_n E[f_n] < \infty$, then $P[E] = 1$, $S(\omega)$ is integrable, and $E[S(\omega)] = \sum_n E[f_n(\omega)]$.

PROOF: By the previous exercise if $\sum_n P(A_n) < \infty$, then $\sum_n \chi_{A_n}(\omega) = S(\omega)$ is finite almost everywhere and

$$E(S(\omega)) = \sum_n P(A_n) < \infty.$$

If an infinite series has a finite sum, then the nth term must go to 0, thereby proving the direct part. To prove the converse we need to show that if $\sum_n P(A_n) = \infty$,

3.3. STRONG LIMIT THEOREMS

then $\lim_{m\to\infty} P(\bigcup_{n=m}^{\infty} A_n) > 0$. We can use independence and the continuity of probability under monotone limits, to calculate for every m,

$$P\left(\bigcup_{n=m}^{\infty} A_n\right) = 1 - P\left(\bigcap_{n=m}^{\infty} A_n^c\right)$$

$$= 1 - \prod_{n=m}^{\infty}(1 - P(A_n)) \quad \text{(by independence)}$$

$$\geq 1 - e^{-\sum_m^{\infty} P(A_n)} > 0$$

and we are done. We have used the inequality $1 - x \leq e^{-x}$ familiar in the study of infinite products. $\square$

Another digression that we want to make into measure theory at this point is to discuss Kolmogorov's consistency theorem. How do we know that there are probability spaces that admit a sequence of independent identically distributed random variables with specified distributions? By the construction of product measures that we outlined earlier we can construct a measure on $\mathbb{R}^n$ for every n which is the joint distribution of the first n random variables. Let us denote by P_n this probability measure on $\mathbb{R}^n$. They are consistent in the sense that if we project in the natural way from $\mathbb{R}^{n+1} \to \mathbb{R}^n$, P_{n+1} projects to P_n. Such a family is called a consistent family of finite-dimensional distributions. We look at the space $\Omega = \mathbb{R}^\infty$ consisting of all real sequences $\omega = \{x_n : n \geq 1\}$ with a natural σ-field Σ generated by the field $\mathcal{F}$ of finite-dimensional cylinder sets of the form $B = \{\omega : (x_1, x_2, \ldots, x_n) \in A\}$ where A varies over Borel sets in $\mathbb{R}^n$ and varies over positive integers.

- **THEOREM 3.5** (Kolmogorov's Consistency Theorem) *Given a consistent family of finite-dimensional distributions P_n, there exists a unique P on (Ω, Σ) such that for every n, under the natural projection $\pi_n(\omega) = (x_1, x_2, \ldots, x_n)$, the induced measure $P\pi_n^{-1} = P_n$ on $\mathbb{R}^n$.*

PROOF: The consistency is just what is required to be able to define P on $\mathcal{F}$ by

$$P(B) = P_n(A).$$

Once we have P defined on the field $\mathcal{F}$, we have to prove the countable additivity of P on $\mathcal{F}$. The rest is then routine. Let $B_n \in \mathcal{F}$ and $B_n \downarrow \emptyset$, the empty set. If possible let $P(B_n) \geq \delta$ for all n and for some $\delta > 0$. Then $B_n = \pi_{k_n}^{-1} A_{k_n}$ for some k_n, and without loss of generality we assume that $k_n = n$, so that $B_n = \pi_n^{-1} A_n$ for some Borel set $A_n \subset \mathbb{R}^n$. According to Exercise 3.9 below, we can find a closed bounded subset $K_n \subset A_n$ such that

$$P_n(A_n - K_n) \leq \frac{\delta}{2^{n+1}}$$

and define $C_n = \pi_n^{-1} K_n$ and $D_n = \bigcap_{j=1}^{n} C_j = \pi_n^{-1} F_n$ for some closed bounded set $F_n \subset K_n \subset \mathbb{R}^n$. Then

$$P(D_n) \geq \delta - \sum_{j=1}^{n} \frac{\delta}{2^{j+1}} \geq \frac{\delta}{2}.$$

$D_n \subset B_n$, $D_n \downarrow \varnothing$, and each D_n is nonempty. If we take $\omega^{(n)} = \{x_j^n : j \geq 1\}$ to be an arbitrary point from D_n, by our construction $(x_1^n, x_2^n, \ldots, x_m^n) \in F_m$ for $n \geq m$. We can definitely choose a subsequence (diagonalization) such that $x_j^{n_k}$ converges for each j producing a limit $\omega = (x_1, x_2, \ldots, x_m, \ldots)$ and, for every m, we will have $(x_1, x_2, \ldots, x_m) \in F_m$. This implies that $\omega \in D_m$ for every m, contradicting $D_n \downarrow \varnothing$. We are done. $\square$

EXERCISE 3.9. We have used the fact that given any Borel set $A \subset \mathbb{R}^n$ and a probability measure α on $\mathbb{R}^n$, for any $\varepsilon > 0$, there exists a closed bounded subset $K_\varepsilon \subset A$ such that $\alpha(A - K_\varepsilon) \leq \varepsilon$. Prove it by showing that the class $\mathcal{A}$ of sets A with the above property is a monotone class that contains finite disjoint unions of measurable rectangles and therefore contains the Borel σ-field. To prove the last fact, establish it first for $n = 1$. To handle $n = 1$, repeat the same argument starting from finite disjoint unions of right-closed left-open intervals. Use the countable additivity to verify this directly.

REMARK 3.2. Kolmogorov's consistency theorem remains valid if we replace $\mathbb{R}$ by an arbitrary complete separable metric space X, with its Borel σ-field. However, it is not valid in complete generality; see [8] and Remark 4.7 in this context.

The following is a strong version of the law of large numbers:

THEOREM 3.6 *If $X_1, X_2, \ldots, X_n, \ldots$ is a sequence of independent identically distributed random variables with $E|X_i|^4 = C < \infty$, then*

$$\lim_{n \to \infty} \frac{S_n}{n} = \lim_{n \to \infty} \frac{X_1 + X_2 + \cdots + X_n}{n} = E(X_1)$$

with probability 1.

PROOF: We can assume without loss of generality that $E[X_i] = 0$; just take $Y_i = X_i - E[X_i]$. A simple calculation shows

$$E[(S_n)^4] = nE[(X_1)^4] + 3n(n-1)E[(X_1)^2]^2 \leq nC + 3n^2\sigma^4,$$

and by applying a Chebyshev-type inequality using fourth moments,

$$P\left[\left|\frac{S_n}{n}\right| \geq \delta\right] = P[|S_n| \geq n\delta] \leq \frac{nC + 3n^2\sigma^4}{n^4\delta^4}.$$

We see that

$$\sum_{n=1}^{\infty} P\left[\left|\frac{S_n}{n}\right| \geq \delta\right] < \infty,$$

and we can now apply the Borel-Cantelli lemma. $\square$

3.4. Series of Independent Random Variables

We wish to investigate conditions under which an infinite series with independent summands

$$S = \sum_{j=1}^{\infty} X_j$$

converges with probability 1. The basic steps are the following inequalities due to Kolmogorov and Lévy that control the behavior of sums of independent random variables. They both deal with the problem of estimating

$$T_n(\omega) = \sup_{1 \le k \le n} |S_k(\omega)| = \sup_{1 \le k \le n} \left| \sum_{j=1}^{k} X_j(\omega) \right|$$

where $X_1, X_2, \ldots, X_n$ are n independent random variables.

LEMMA 3.7 (Kolmogorov's Inequality) *Assume that $EX_i = 0$ and $\text{Var}(X_i) = \sigma_i^2 < \infty$ and let $s_n^2 = \sum_{j=1}^{n} \sigma_j^2$; then*

(3.9)
$$P\{T_n(\omega) \ge \ell\} \le \frac{s_n^2}{\ell^2}.$$

PROOF: The important point here is that the estimate depends only on s_n^2 and not on the number of summands. In fact, the Chebyshev bound on S_n is

$$P\{|S_n| \ge \ell\} \le \frac{s_n^2}{\ell^2}$$

and the supremum does not cost anything.

Let us define the events $E_k = \{|S_1| < \ell, \ldots, |S_{k-1}| < \ell, |S_k| \ge \ell\}$; then $\{T_n \ge \ell\} = \bigcup_{k=1}^{n} E_k$ is a disjoint union of E_k. If we use the independence of $S_n - S_k$ and $S_k \chi_{E_k}$ that depends only on $X_1, X_2, \ldots, X_k$,

$$P\{E_k\} \le \frac{1}{\ell^2} \int_{E_k} S_k^2 \, dP \le \frac{1}{\ell^2} \int_{E_k} [S_k^2 + (S_n - S_k)^2] \, dP$$

$$= \frac{1}{\ell^2} \int_{E_k} [S_k^2 + 2S_k(S_n - S_k) + (S_n - S_k)^2] \, dP$$

$$= \frac{1}{\ell^2} \int_{E_k} S_n^2 \, dP.$$

Summing over k from 1 to n

$$P\{T_n \ge \ell\} \le \frac{1}{\ell^2} \int_{T_n \ge \ell} S_n^2 \, dP \le \frac{s_n^2}{\ell^2},$$

establishing (3.9). □

LEMMA 3.8 (Lévy's Inequality) *Assume that*

$$P\left\{|X_i + X_{i+1} + \cdots + X_n| \geq \frac{\ell}{2}\right\} \leq \delta \quad \text{for all } 1 \leq i \leq n;$$

then

(3.10)
$$P\{T_n \geq \ell\} \leq \frac{\delta}{1-\delta}.$$

PROOF: Let E_k be as in the previous lemma.

$$P\left\{(T_n \geq \ell) \cap |S_n| \leq \frac{\ell}{2}\right\} = \sum_{k=1}^n P\left\{E_k \cap |S_n| \leq \frac{\ell}{2}\right\}$$

$$\leq \sum_{k=1}^n P\left\{E_k \cap |S_n - S_k| \geq \frac{\ell}{2}\right\}$$

$$= \sum_{k=1}^n P\left\{|S_n - S_k| \geq \frac{\ell}{2}\right\} P(E_k)$$

$$\leq \delta \sum_{k=1}^n P(E_k) = \delta P\{T_n \geq \ell\}.$$

On the other hand,

$$P\left\{(T_n \geq \ell) \cap |S_n| > \frac{\ell}{2}\right\} \leq P\left\{|S_n| > \frac{\ell}{2}\right\} \leq \delta.$$

Adding the two,

$$P\{T_n \geq \ell\} \leq \delta P\{T_n \geq \ell\} + \delta \quad \text{or} \quad P\{T_n \geq \ell\} \leq \frac{\delta}{1-\delta},$$

proving (3.10). □

We are now ready to prove

THEOREM 3.9 (Lévy's Theorem) *If* $X_1, X_2, \ldots, X_n, \ldots$ *is a sequence of independent random variables, then the following are equivalent*:
  (i) *The distribution α_n of $S_n = X_1 + X_2 + \cdots + X_n$ converges weakly to a probability distribution α on $\mathbb{R}$.*
  (ii) *The random variable $S_n = X_1 + X_2 + \cdots + X_n$ converges in probability to a limit $S(\omega)$.*
  (iii) *The random variable $S_n = X_1 + X_2 + \cdots + X_n$ converges with probability 1 to a limit $S(\omega)$.*

PROOF: Clearly (iii) $\Rightarrow$ (ii) $\Rightarrow$ (i) are trivial. We will establish (i) $\Rightarrow$ (ii) $\Rightarrow$ (iii).

(i) $\Rightarrow$ (ii): The characteristic functions $\phi_j(t)$ of X_j are such that

$$\phi(t) = \prod_{i=1}^\infty \phi_j(t)$$

3.4. SERIES OF INDEPENDENT RANDOM VARIABLES

is a convergent infinite product. Since the limit $\phi(t)$ is continuous at $t = 0$ and $\phi(0) = 1$, it is nonzero in some interval $|t| \leq T$ around 0. Therefore, for $|t| \leq T$,

$$\lim_{\substack{n\to\infty \\ m\to\infty}} \prod_{m+1}^{n} \phi_j(t) = 1. \quad (tail)$$

By Exercise 3.10 below, this implies that for all t,

$$\lim_{\substack{n\to\infty \\ m\to\infty}} \prod_{m+1}^{n} \phi_j(t) = 1$$

and, consequently, the distribution of $S_n - S_m$ converges to the distribution degenerate at 0. This implies the convergence in probability to 0 of $S_n - S_m$ as $m, n \to \infty$. Therefore for each $\delta > 0$,

$$\lim_{\substack{n\to\infty \\ m\to\infty}} P\{|S_n - S_m| \geq \delta\} = 0$$

establishing (ii).

(ii) ⇒ (iii): To establish (iii), because of Exercise 3.11 below, we need only show that for every $\delta > 0$

$$\lim_{\substack{n\to\infty \\ m\to\infty}} P\left\{\sup_{m<k\leq n} |S_k - S_m| \geq \delta\right\} = 0,$$

and this follows from (ii) and Lévy's inequality. □

EXERCISE 3.10. Prove the inequality $1 - \cos 2t \leq 4(1 - \cos t)$ for all real t. Deduce the inequality $1 - \mathrm{Re}\,\phi(2t) \leq 4[1 - \mathrm{Re}\,\phi(t)]$, valid for any characteristic function. Conclude that if a sequence of characteristic functions converges to 1 in an interval around 0, then it converges to 1 for all real t.

EXERCISE 3.11. Prove that if a sequence S_n of random variables is a Cauchy sequence in probability, i.e.,

$$\lim_{\substack{n\to\infty \\ m\to\infty}} P\{|S_n - S_m| \geq \delta\} = 0 \quad \text{for each } \delta > 0,$$

then there is a random variable S such that $S_n \to S$ in probability, i.e.,

$$\lim_{n\to\infty} P\{|S_n - S| \geq \delta\} = 0 \quad \text{for each } \delta > 0.$$

EXERCISE 3.12. Prove that if a sequence S_n of random variables satisfies

$$\lim_{\substack{n\to\infty \\ m\to\infty}} P\left\{\sup_{m<k\leq n} |S_k - S_m| \geq \delta\right\} = 0 \quad \text{for every } \delta > 0,$$

then there is a limiting random variable $S(\omega)$ such that

$$P\left\{\lim_{n\to\infty} S_n(\omega) = S(\omega)\right\} = 1.$$

EXERCISE 3.13. Prove that whenever $X_n \to X$ in probability, the distribution α_n of X_n converges weakly to the distribution α of X.

Now it is straightforward to find sufficient conditions for the convergence of an infinite series of <u>independent</u> random variables.

- THEOREM 3.10 (Kolmogorov's One Series Theorem) *Let a sequence $\{X_i\}$ of independent random variables, each of which has <u>two moments</u>, satisfy $E(X_i) = 0$ and $\sum_{i=1}^{\infty} \operatorname{Var}(X_i) < \infty$; then*

$$S(\omega) = \sum_{i=1}^{\infty} X_i(\omega)$$

<u>*converges with probability* 1</u>.

PROOF: By a direct application of Kolmogorov's inequality

$$\lim_{\substack{n\to\infty\\m\to\infty}} P\left\{\sup_{m<k\leq n} |S_k - S_m| \geq \delta\right\} \leq \lim_{\substack{n\to\infty\\m\to\infty}} \frac{1}{\delta^2} \sum_{j=m+1}^{n} E(X_j^2)$$

$$= \lim_{\substack{n\to\infty\\m\to\infty}} \frac{1}{\delta^2} \sum_{j=m+1}^{n} \operatorname{Var}(X_i) = 0.$$

Therefore

$$\lim_{\substack{n\to\infty\\m\to\infty}} P\left\{\sup_{m<k\leq n} |S_k - S_m| \geq \delta\right\} = 0.$$

We can also prove <u>convergence in probability</u>

$$\lim_{\substack{n\to\infty\\m\to\infty}} P\{|S_n - S_m| \geq \delta\} = 0$$

by a simple application of Chebyshev's inequality and then apply Lévy's theorem to get almost sure convergence. □

- - THEOREM 3.11 (Kolmogorov's Two Series Theorem) *Let $a_i = E[X_i]$ be the means and $\sigma_i^2 = \operatorname{Var}(X_i)$ the variances of a sequence of independent random variables $\{X_i\}$. Assume that $\sum_i a_i$ and $\sum_i \sigma_i^2$ converge. Then the series $\sum_i X_i$ converges with probability 1.*

PROOF: Define $Y_i = X_i - a_i$ and apply the previous (one series) theorem to Y_i. □

Of course, in general, random variables need not have finite expectations or variances. If $\{X_i\}$ is any sequence of random variables, we can take a cutoff value C and define $Y_i = X_i$ if $|X_i| \leq C$ and $Y_i = 0$ otherwise. The Y_i are then independent and bounded in absolute value by C. The theorem can be applied to Y_i and if we impose <u>the additional condition</u> that

$$\sum_i P\{X_i \neq Y_i\} = \sum_i P\{|X_i| > C\} < \infty$$

by an application of the Borel-Cantelli lemma, with probability 1, <u>$X_i = Y_i$ for all sufficiently large i</u>. The convergence of $\sum_i X_i$ and $\sum_i Y_i$ are therefore equivalent. We get then the sufficiency part of

3.4. SERIES OF INDEPENDENT RANDOM VARIABLES

♦♦♦ THEOREM 3.12 (Kolmogorov's Three Series Theorem) *For the convergence of an infinite series of independent random variables $\sum_i X_i$, it is necessary and sufficient that all the three following infinite series converge*: *— sort of tightness*
  (i) *For some cutoff value $C > 0$, $\sum_i P\{|X_i| > C\}$ converges.*
  (ii) *If Y_i is defined to equal X_i if $|X_i| \leq C$ and 0 otherwise, $\sum_i E(Y_i)$ converges.*
  (iii) *With Y_i as in (ii), $\sum_i \mathrm{Var}(Y_i)$ converges.* *(truncated var conv.)*

PROOF: Let us now prove the converse. If $\sum_i X_i$ converges for a sequence of independent random variables, we must necessarily have $|X_n| \leq C$ eventually with probability 1. By the Borel-Cantelli lemma the first series must converge. This means that in order to prove the necessity we can assume without loss of generality that the $|X_i|$ are all bounded, say by 1. We may also assume that $E(X_i) = 0$ for each i. Otherwise, let us take independent random variables X'_i that have the same distribution as X_i. Then $\sum_i X_i$ as well as $\sum_i X'_i$ converge with probability 1 and therefore so does $\sum_i (X_i - X'_i)$. The random variables $Z_i = X_i - X'_i$ are independent and bounded by 2. They have mean 0. If we can show $\sum \mathrm{Var}(Z_i)$ is convergent, since $\mathrm{Var}(Z_i) = 2\mathrm{Var}(X_i)$ we will prove the convergence of the third series. Now it is elementary to conclude that since both $\sum_i X_i$ as well as $\sum_i (X_i - E(X_i))$ converge, the series $\sum_i E(X_i)$ must be convergent as well. □

All we need is the following lemma to complete the proof of necessity:

LEMMA 3.13 *If $\sum_i X_i$ is convergent for a series of independent random variables with mean 0 that are individually bounded by C, then $\sum_i \mathrm{Var}(X_i)$ is convergent.*

PROOF: Let $F_n = \{\omega : |S_1| \leq \ell, |S_2| \leq \ell, \ldots, |S_n| \leq \ell\}$ where $S_k = X_1 + X_2 + \cdots + X_k$. If the series converges with probability 1, we must have, for some ℓ and $\delta > 0$, $P(F_n) \geq \delta$ for all n. We have

$$\int_{F_{n-1}} S_n^2 \, dP = \int_{F_{n-1}} [S_{n-1} + X_n]^2 \, dP$$

$$= \int_{F_{n-1}} \left[S_{n-1}^2 + 2 S_{n-1} X_n + X_n^2 \right] dP$$

$$= \int_{F_{n-1}} S_{n-1}^2 \, dP + \sigma_n^2 P(F_{n-1}) \geq \int_{F_{n-1}} S_{n-1}^2 \, dP + \delta \sigma_n^2$$

and, on the other hand,

$$\int_{F_{n-1}} S_n^2 \, dP = \int_{F_n} S_n^2 \, dP + \int_{F_{n-1} \cap F_n^c} S_n^2 \, dP \leq \int_{F_n} S_n^2 \, dP + P(F_{n-1} \cap F_n^c)(\ell + C)^2 ,$$

providing us with the estimate

$$\delta \sigma_n^2 \leq \int_{F_n} S_n^2 \, dP - \int_{F_{n-1}} S_{n-1}^2 \, dP + P(F_{n-1} \cap F_n^c)(\ell + C)^2 .$$

Since $F_{n-1} \cap F_n^c$ are disjoint and $|S_n| \leq \ell$ on F_n,

$$\sum_{j=1}^{\infty} \sigma_j^2 \leq \frac{1}{\delta}[\ell^2 + (\ell + C)^2].$$

This concludes the proof. □

3.5. Strong Law of Large Numbers

We saw earlier in Theorem 3.6 that if $\{X_i\}$ is a sequence of i.i.d. (independent identically distributed) random variables with zero mean and a finite fourth moment, then $\frac{X_1+X_2+\cdots+X_n}{n} \to 0$ with probability 1. We will now prove the same result assuming only that $E|X_i| < \infty$ and $E(X_i) = 0$.

THEOREM 3.14 *If $\{X_i\}$ is a sequence of i.i.d. random variables with mean 0,*

$$\lim_{n\to\infty} \frac{X_1 + X_2 + \cdots + X_n}{n} = 0$$

with probability 1.

PROOF: We define

$$Y_n = \begin{cases} X_n & \text{if } |X_n| \leq n \\ 0 & \text{if } |X_n| > n \end{cases}$$

$$a_n = P[X_n \neq Y_n], \quad b_n = E[Y_n], \quad \text{and} \quad c_n = \text{Var}(Y_n).$$

First we note that (see exercise 3.14 below)

$$\sum_n a_n = \sum_n P[|X_1| > n] \leq E|X_1| < \infty, \quad \lim_{n\to\infty} b_n = 0,$$

and

$$\sum_n \frac{c_n}{n^2} \leq \sum_n \frac{E[Y_n^2]}{n^2} = \sum_n \int_{|x|\leq n} \frac{x^2}{n^2} d\alpha$$

$$= \int x^2 \left(\sum_{n\geq x} \frac{1}{n^2} \right) d\alpha \leq C \int |x| d\alpha < \infty$$

where α is the common distribution of X_i. From the three series theorem and the Borel-Cantelli lemma, we conclude that $\sum_n \frac{Y_n-b_n}{n}$ as well as $\sum_n \frac{X_n-b_n}{n}$ converge almost surely. It is elementary to verify that for any series $\sum_n \frac{x_n}{n}$ that converges, $\frac{x_1+x_2+\cdots+x_n}{n} \to 0$ as $n \to \infty$. We therefore conclude that

$$P\left\{ \lim_{n\to\infty} \left[\frac{X_1 + X_2 + \cdots + X_n}{n} - \frac{b_1 + b_2 + \cdots + b_n}{n} \right] = 0 \right\} = 1.$$

Since $b_n \to 0$ as $n \to \infty$, the theorem is proved. □

3.6. CENTRAL LIMIT THEOREM

EXERCISE 3.14. Let X be a nonnegative random variable. Then

$$E[X] - 1 \leq \sum_{n=1}^{\infty} P[X \geq n] \leq E[X].$$

In particular, $E[X] < \infty$ if and only if $\sum_n P[X \geq n] < \infty$.

EXERCISE 3.15 If the strong law of large numbers holds for a sequence of i.i.d. random variables $X_1, X_2, \ldots, X_n, \ldots$, with some limit, i.e.,

$$P\left[\lim_{n \to \infty} \frac{S_n}{n} = \xi\right] = 1 \quad \text{for some random variable } \xi,$$

which may or may not be a constant with probability 1, then show that necessarily $E|X_i| < \infty$. Consequently, $\xi = E(X_i)$ with probability 1.

One may ask why the limit cannot be a proper random variable. There is a general theorem that forbids it called *Kolmogorov's zero-one law*. Let us look at the space Ω of real sequences $\{x_n : n \geq 1\}$. We have the σ-field $\mathcal{B}$, the product σ-field on Ω. In addition, we have the sub σ-fields $\mathcal{B}^n$ generated by $\{x_j : j \geq n\}$. $\mathcal{B}^n$ are $\downarrow$ with n and $\mathcal{B}^\infty = \bigcap_n \mathcal{B}^n$, which is also a σ-field and is called the *tail* σ-field. The typical set in $\mathcal{B}^\infty$ is a set depending only on the tail behavior of the sequence. For example, the sets $\{\omega : x_n \text{ is bounded}\}$ and $\{\omega : \limsup_n x_n = 1\}$ are in $\mathcal{B}^\infty$, whereas $\{\omega : \sup_n |x_n| = 1\}$ is not.

THEOREM 3.15 (Kolmogorov's Zero-One Law) *If $A \in \mathcal{B}^\infty$ and P is any product measure (not necessarily with identical components), $P(A) = 0$ or 1.*

PROOF: The proof depends on showing that A is independent of itself so that $P(A) = P(A \cap A) = P(A)P(A) = [P(A)]^2$ and therefore equals 0 or 1. The proof is elementary. Since $A \in \mathcal{B}^\infty \subset \mathcal{B}^{n+1}$ and P is a product measure, A is independent of $\mathcal{B}_n = \sigma$-field generated by $\{x_j : 1 \leq j \leq n\}$. It is, therefore, independent of sets in the field $\mathcal{F} = \bigcup_n \mathcal{B}_n$. The class $\mathcal{A}$ of sets that are independent of A is a monotone class. Since it contains the field $\mathcal{F}$, it contains the σ-field $\mathcal{B}$ generated by $\mathcal{F}$. In particular, since $A \in \mathcal{B}$, A is independent of itself. □

COROLLARY 3.16 *Any random variable measurable with respect to the tail σ-field $\mathcal{B}^\infty$ is equal with probability 1 to a constant relative to any given product measure.*

PROOF: Left as an exercise. □

 Warning. For different product measures the constants can be different.

EXERCISE 3.16 How can that happen?

3.6. Central Limit Theorem

We saw before that for any sequence of independent identically distributed random variables $X_1, X_2, \ldots, X_n, \ldots$ the sum $S_n = X_1 + X_2 + \cdots + X_n$ has the property that

$$\lim_{n \to \infty} \frac{S_n}{n} = 0$$

in probability provided the expectation exists and equals 0. If we assume that the variance of the random variables is finite and equals $\sigma^2 > 0$, then we have

THEOREM 3.17 *The distribution of $\frac{S_n}{\sqrt{n}}$ converges as $n \to \infty$ to the normal distribution with density*

(3.11) $$p(x) = \frac{1}{\sqrt{2\pi}\sigma} \exp\left[-\frac{x^2}{2\sigma^2}\right].$$

PROOF: If we denote by $\phi(t)$ the characteristic function of any X_i, then the characteristic function of $\frac{S_n}{\sqrt{n}}$ is given by

$$\psi_n(t) = \left[\phi\left(\frac{t}{\sqrt{n}}\right)\right]^n.$$

We can use the expansion

$$\phi(t) = 1 - \frac{\sigma^2 t^2}{2} + o(t^2)$$

to conclude that

$$\phi\left(\frac{t}{\sqrt{n}}\right) = 1 - \frac{\sigma^2 t^2}{2n} + o\left(\frac{1}{n}\right),$$

and it then follows that

$$\lim_{n\to\infty} \psi_n(t) = \psi(t) = \exp\left[-\frac{\sigma^2 t^2}{2}\right].$$

Since $\psi(t)$ is the characteristic function of the normal distribution with density $p(x)$ given by equation (3.11), we are done. $\square$

EXERCISE 3.17. A more direct proof is possible in some special cases. For instance, if each $X_i = \pm 1$ with probability $\frac{1}{2}$, S_n can take the values $n - 2k$ with $0 \le k \le n$,

$$P[S_n = 2k - n] = \frac{1}{2^n}\binom{n}{k} \quad \text{and} \quad P\left[a \le \frac{S_n}{\sqrt{n}} \le b\right] = \frac{1}{2^n} \sum_{k: a\sqrt{n} \le 2k-n \le b\sqrt{n}} \binom{n}{k}.$$

Use Stirling's formula to prove directly that

$$\lim_{n\to\infty} P\left[a \le \frac{S_n}{\sqrt{n}} \le b\right] = \int_a^b \frac{1}{\sqrt{2\pi}} \exp\left[-\frac{x^2}{2}\right] dx.$$

Actually, for the proof of the central limit theorem we do not need the random variables $\{X_j\}$ to have identical distributions. Let us suppose that they all have zero means and that the variance of X_j is σ_j^2. Define $s_n^2 = \sigma_1^2 + \sigma_2^2 + \cdots + \sigma_n^2$. Assume $s_n^2 \to \infty$ as $n \to \infty$. Then $Y_n = \frac{S_n}{s_n}$ has zero mean and unit variance. It is not unreasonable to expect that

$$\lim_{n\to\infty} P[Y_n \le a] = \int_{-\infty}^a \frac{1}{\sqrt{2\pi}} \exp\left[-\frac{x^2}{2}\right] dx$$

under certain mild conditions.

3.6. CENTRAL LIMIT THEOREM

THEOREM 3.18 (Lindeberg's Theorem) *If we denote by α_i the distribution of X_i, the condition (known as Lindeberg's condition)*

$$\lim_{n\to\infty} \frac{1}{s_n^2} \sum_{i=1}^{n} \int_{|x|\geq \varepsilon s_n} x^2 \, d\alpha_i = 0 \quad \text{for each } \varepsilon > 0$$

is sufficient for the central limit theorem to hold.

PROOF: The first step in proving this limit theorem as well as other limit theorems that we will prove is to rewrite

$$Y_n = X_{n,1} + X_{n,2} + \cdots + X_{n,k_n} + A_n$$

where $X_{n,j}$ are k_n mutually independent random variables and A_n is a constant. In our case $k_n = n$, $A_n = 0$, and $X_{n,j} = \frac{X_j}{s_n}$ for $1 \leq j \leq n$. We denote by

$$\phi_{n,j}(t) = E[e^{it X_{n,j}}] = \int e^{itx} \, d\alpha_{n,j} = \int e^{it\frac{x}{s_n}} \, d\alpha_j = \phi_j\left(\frac{t}{s_n}\right)$$

where $\alpha_{n,j}$ is the distribution of $X_{n,j}$. The functions ϕ_j and $\phi_{n,j}$ are the characteristic functions of α_j and $\alpha_{n,j}$, respectively. If we denote by μ_n the distribution of Y_n, its characteristic function $\hat{\mu}_n(t)$ is given by

$$\hat{\mu}_n(t) = \prod_{j=1}^{n} \phi_{n,j}(t)$$

and our goal is to show that

$$\lim_{n\to\infty} \hat{\mu}_n(t) = \exp\left[-\frac{t^2}{2}\right].$$

This will be carried out in several steps. First, we define

$$\psi_{n,j}(t) = \exp[\phi_{n,j}(t) - 1] \quad \text{and} \quad \psi_n(t) = \prod_{j=1}^{n} \psi_{n,j}(t).$$

We show that for each finite T,

$$\lim_{n\to\infty} \sup_{|t|\leq T} \sup_{1\leq j\leq n} |\phi_{n,j}(t) - 1| = 0 \quad \text{and} \quad \sup_n \sup_{|t|\leq T} \sum_{j=1}^{n} |\phi_{n,j}(t) - 1| < \infty.$$

This would imply that

$$\lim_{n\to\infty} \sup_{|t|\leq T} |\log \hat{\mu}_n(t) - \log \psi_n(t)|$$

$$\leq \lim_{n\to\infty} \sup_{|t|\leq T} \sum_{j=1}^{n} |\log \phi_{n,j}(t) - [\phi_{n,j}(t) - 1]|$$

$$\le \lim_{n\to\infty} \sup_{|t|\le T} C \sum_{j=1}^{n} |\phi_{n,j}(t) - 1|^2$$

$$\le C \lim_{n\to\infty} \left\{ \sup_{|t|\le T} \sup_{1\le j\le n} |\phi_{n,j}(t) - 1| \right\} \left\{ \sup_{|t|\le T} \sum_{j=1}^{n} |\phi_{n,j}(t) - 1| \right\} = 0$$

by the expansion

$$\log r = \log(1 + (r-1)) = r - 1 + O(r-1)^2.$$

The proof can then be completed by showing

$$\lim_{n\to\infty} \sup_{|t|\le T} \left| \log \psi_n(t) + \frac{t^2}{2} \right| = \lim_{n\to\infty} \sup_{|t|\le T} \left| \left[\sum_{j=1}^{n} (\phi_{n,j}(t) - 1) \right] + \frac{t^2}{2} \right| = 0.$$

We see that

$$\sup_{|t|\le T} |\phi_{n,j}(t) - 1| = \sup_{|t|\le T} \left| \int [\exp[itx] - 1] d\alpha_{n,j} \right|$$

$$= \sup_{|t|\le T} \left| \int \left[\exp\left[it\frac{x}{s_n} \right] - 1 \right] d\alpha_j \right|$$

(3.12)
$$= \sup_{|t|\le T} \left| \int \left[\exp\left[it\frac{x}{s_n} \right] - 1 - it\frac{x}{s_n} \right] d\alpha_j \right|$$

(3.13)
$$\le C_T \int \frac{x^2}{s_n^2} d\alpha_j$$

$$= C_T \int_{|x|<\varepsilon s_n} \frac{x^2}{s_n^2} d\alpha_j + C_T \int_{|x|\ge\varepsilon s_n} \frac{x^2}{s_n^2} d\alpha_j$$

(3.14)
$$\le C_T \varepsilon^2 + C_T \frac{1}{s_n^2} \int_{|x|\ge\varepsilon s_n} x^2 d\alpha_j.$$

We have used the mean zero condition in deriving equation (3.12) and the estimate $|e^{ix} - 1 - ix| \le cx^2$ to get to the equation (3.13). If we let $n \to \infty$, by Lindeberg's condition, the second term of equation (3.14) goes to 0. Therefore

$$\limsup_{n\to\infty} \sup_{1\le j\le k_n} \sup_{|t|\le T} |\phi_{n,j}(t) - 1| \le \varepsilon^2 C_T.$$

Since $\varepsilon > 0$ is arbitrary, we have

$$\lim_{n\to\infty} \sup_{1\le j\le k_n} \sup_{|t|\le T} |\phi_{n,j}(t) - 1| = 0.$$

Next we observe that there is a bound,

$$\sup_{|t|\le T} \sum_{j=1}^{n} |\phi_{n,j}(t) - 1| \le C_T \sum_{j=1}^{n} \int \frac{x^2}{s_n^2} d\alpha_j \le C_T \frac{1}{s_n^2} \sum_{j=1}^{n} \sigma_j^2 = C_T$$

uniformly in n. Finally, for each $\varepsilon > 0$,

$$\lim_{n\to\infty} \sup_{|t|\leq T} \left| \left[\sum_{j=1}^{n} (\phi_{n,j}(t) - 1) \right] + \frac{t^2}{2} \right|$$

$$\leq \lim_{n\to\infty} \sup_{|t|\leq T} \sum_{j=1}^{n} \left| \phi_{n,j}(t) - 1 + \frac{\sigma_j^2 t^2}{2 s_n^2} \right|$$

$$= \lim_{n\to\infty} \sup_{|t|\leq T} \sum_{j=1}^{n} \left| \int \left[\exp\left[it\frac{x}{s_n}\right] - 1 - it\frac{x}{s_n} + \frac{t^2 x^2}{2s_n^2} \right] d\alpha_j \right|$$

$$\leq \lim_{n\to\infty} \sup_{|t|\leq T} \sum_{j=1}^{n} \left| \int_{|x|<\varepsilon s_n} \left[\exp\left[it\frac{x}{s_n}\right] - 1 - it\frac{x}{s_n} + \frac{t^2 x^2}{2s_n^2} \right] d\alpha_j \right|$$

$$+ \lim_{n\to\infty} \sup_{|t|\leq T} \sum_{j=1}^{n} \left| \int_{|x|\geq\varepsilon s_n} \left[\exp\left[it\frac{x}{s_n}\right] - 1 - it\frac{x}{s_n} + \frac{t^2 x^2}{2s_n^2} \right] d\alpha_j \right|$$

$$\leq \lim_{n\to\infty} C_T \sum_{j=1}^{n} \int_{|x|<\varepsilon s_n} \frac{|x|^3}{s_n^3} d\alpha_j + \lim_{n\to\infty} C_T \sum_{j=1}^{n} \int_{|x|\geq\varepsilon s_n} \frac{x^2}{s_n^2} d\alpha_j$$

$$\leq \varepsilon C_T \lim_{n\to\infty} \sup \sum_{j=1}^{n} \int \frac{x^2}{s_n^2} d\alpha_j + \lim_{n\to\infty} C_T \sum_{j=1}^{n} \int_{|x|\geq\varepsilon s_n} \frac{x^2}{s_n^2} d\alpha_j = \varepsilon C_T$$

by Lindeberg's condition. Since $\varepsilon > 0$ is arbitrary, the result is proven. □

REMARK 3.3. The key step in the proof of the central limit theorem under Lindeberg's condition, as well as in other limit theorems for sums of independent random variables, is the analysis of products

$$\psi_n(t) = \prod_{j=1}^{k_n} \phi_{n,j}(t).$$

The idea is to replace each $\phi_{n,j}(t)$ by $\exp[\phi_{n,j}(t) - 1]$, changing the product to the exponential of a sum. Although each $\phi_{n,j}(t)$ is close to 1, making the idea reasonable, in order for the idea to work one has to show that the sum $\sum_{j=1}^{k_n} |\phi_{n,j}(t) - 1|^2$ is negligible. This requires the boundedness of $\sum_{j=1}^{k_n} |\phi_{n,j}(t) - 1|$. One has to use the mean zero condition or some suitable centering condition to cancel the first term in the expansion of $\phi_{n,j}(t) - 1$ and control the rest from sums of the variances.

EXERCISE 3.18. Lyapunov's condition is the following: For some $\delta > 0$

$$\lim_{n\to\infty} \frac{1}{s_n^{2+\delta}} \sum_{j=1}^{n} \int |x|^{2+\delta} d\alpha_j = 0.$$

Prove that Lyapunov's condition implies Lindeberg's condition.

EXERCISE 3.19. Consider the case of mutually independent random variables $\{X_j\}$, where $X_j = \pm a_j$ with probability $\frac{1}{2}$. What do Lyapunov's and Lindeberg's conditions demand of $\{a_j\}$? Can you find a sequence $\{a_j\}$ that does not satisfy Lyapunov's condition for any $\delta > 0$ but satisfies Lindeberg's condition? Try to find a sequence $\{a_j\}$ such that the central limit theorem is not valid.

3.7. Accompanying Laws

As we stated in the previous section, we want to study the behavior of the sum of a large number of independent random variables. We have k_n independent random variables $\{X_{n,j} : 1 \leq j \leq k_n\}$ with respective distributions $\{\alpha_{n,j}\}$. We are interested in the distribution μ_n of $Z_n = \sum_{j=1}^{k_n} X_{n,j}$. One important assumption that we will make on the random variables $\{X_{n,j}\}$ is that <u>no single one is significant</u>. More precisely, for every $\delta > 0$,

(3.15) $$\lim_{n \to \infty} \sup_{1 \leq j \leq k_n} P[|X_{n,j}| \geq \delta] = \lim_{n \to \infty} \sup_{1 \leq j \leq k_n} \alpha_{n,j}[|x| \geq \delta] = 0.$$

The condition is referred to as *uniform infinitesimality*. The following construction will play a major role: If α is a probability distribution on the line and $\phi(t)$ is its characteristic function, for any nonnegative real number $a > 0$, $\psi_a(t) = \exp[a(\phi(t) - 1)]$ is again a characteristic distribution. In fact, if we denote by α^k the k-fold convolution of α with itself, ψ_a is seen to be the characteristic function of the probability distribution

$$e^{-a} \sum_{j=0}^{\infty} \frac{a^j}{j!} \alpha^j,$$

which is a convex combination α^j with weights $e^{-a} \frac{a^j}{j!}$. We use the construction mostly with $a = 1$. If we denote the probability distribution with characteristic function $\psi_a(t)$ by $e_a(\alpha)$, one checks easily that $e_{a+b}(\alpha) = e_a(\alpha) * e_b(\alpha)$. In particular, $e_a(\alpha) = e_{\frac{a}{n}}(\alpha)^n$. Probability distributions β that can be written for each $n \geq 1$ as the n-fold convolution β_n^n of some probability distribution β_n are called *infinitely divisible*. In particular, for every $a \geq 0$ and α, $e_a(\alpha)$ is an infinitely divisible probability distribution. These are called *compound Poisson distributions*. A special case when $\alpha = \delta_1$, the degenerate distribution at 1, we get for $e_a(\delta_1)$ the usual Poisson distribution with parameter a. We can interpret $e_a(\alpha)$ as the distribution of the sum of a random number of independent random variables with common distribution α. The random number n has a distribution that is Poisson with parameter a and is independent of the random variables involved in the sum.

In order to study the distribution μ_n of Z_n, it will be more convenient to replace $\alpha_{n,j}$ by an infinitely divisible distribution $\beta_{n,j}$. This is done as follows: We define

$$a_{n,j} = \int_{|x| \leq 1} x \, d\alpha_{n,j},$$

$\alpha'_{n,j}$ as the translate of $\alpha_{n,j}$ by $-a_{n,j}$, i.e.,

$$\alpha'_{n,j} = \alpha_{n,j} * \delta_{-a_{n,j}}, \quad \beta'_{n,j} = e_1(\alpha'_{n,j}), \quad \beta_{n,j} = \beta'_{n,j} * a_{n,j},$$

3.7. ACCOMPANYING LAWS

and, finally,
$$\lambda_n = \prod_{j=1}^{k_n} \beta_{n,j}.$$

A main tool in this subject is the following theorem. We assume always that the uniform infinitesimality condition (3.15) holds. In terms of notation, we will find it more convenient to denote by $\hat{\mu}$ the characteristic function of the probability distribution μ.

THEOREM 3.19 (Accompanying Laws) *In order that, for some constants A_n, the distribution $\mu_n * \delta_{A_n}$ of $Z_n + A_n$ may converge to the limit μ, it is necessary and sufficient that, for the same constants A_n, the distribution $\lambda_n * \delta_{A_n}$ converges to the same limit μ.*

PROOF: First we note that, for any $\delta > 0$,

$$\limsup_{n\to\infty} \sup_{1\leq j\leq k_n} |a_{n,j}|$$

$$= \limsup_{n\to\infty} \sup_{1\leq j\leq k_n} \left| \int_{|x|\leq 1} x\, d\alpha_{n,j} \right|$$

$$\leq \limsup_{n\to\infty} \sup_{1\leq j\leq k_n} \left| \int_{|x|\leq \delta} x\, d\alpha_{n,j} \right| + \limsup_{n\to\infty} \sup_{1\leq j\leq k_n} \left| \int_{\delta<|x|\leq 1} x\, d\alpha_{n,j} \right|$$

$$\leq \delta + \limsup_{n\to\infty} \sup_{1\leq j\leq k_n} \alpha_{n,j}[|x| \geq \delta] = \delta.$$

Therefore,
$$\lim_{n\to\infty} \sup_{1\leq j\leq k_n} |a_{n,j}| = 0.$$

This means that the $\alpha'_{n,j}$ are uniformly infinitesimal just as the $\alpha_{n,j}$ were. Let us suppose that n is so large that $\sup_{1\leq j\leq k_n} |a_{n,j}| \leq \frac{1}{4}$. The advantage in going from $\alpha_{n,j}$ to $\alpha'_{n,j}$ is that the latter are better centered and we can calculate

$$a'_{n,j} = \int_{|x|\leq 1} x\, d\alpha'_{n,j} = \int_{|x-a_{n,j}|\leq 1} (x - a_{n,j}) d\alpha_{n,j}$$

$$= \int_{|x-a_{n,j}|\leq 1} x\, d\alpha_{n,j} - a_{n,j}\alpha_{n,j}[|x - a_{n,j}| \leq 1]$$

$$= \int_{|x-a_{n,j}|\leq 1} x\, d\alpha_{n,j} - a_{n,j} + a_{n,j}\alpha_{n,j}[|x - a_{n,j}| > 1]$$

and estimate $|a'_{n,j}|$ by

$$|a'_{n,j}| \leq C\alpha_{n,j}\left[|x| \geq \tfrac{3}{4}\right] \leq C\alpha'_{n,j}\left[|x| \geq \tfrac{1}{2}\right].$$

In other words, we may assume without loss of generality that $\alpha_{n,j}$ satisfies the bound

(3.16) $$|a_{n,j}| \leq C\alpha_{n,j}\left[|x| \geq \tfrac{1}{2}\right]$$

and forget all about the change from $\alpha_{n,j}$ to $\alpha'_{n,j}$. We will drop the primes and stay with just $\alpha_{n,j}$. Then, just as in the proof of the Lindeberg theorem, we proceed to estimate

$$\lim_{n\to\infty}\sup_{|t|\leq T}|\log\hat{\lambda}_n(t) - \log\hat{\mu}_n(t)|$$

$$\leq \lim_{n\to\infty}\sup_{|t|\leq T}\left|\sum_{j=1}^{k_n}[\log\hat{\alpha}_{n,j}(t) - (\hat{\alpha}_{n,j}(t)-1)]\right|$$

$$\leq \lim_{n\to\infty}\sup_{|t|\leq T}\sum_{j=1}^{k_n}|\log\hat{\alpha}_{n,j}(t) - (\hat{\alpha}_{n,j}(t)-1)|$$

$$\leq \lim_{n\to\infty}\sup_{|t|\leq T}C\sum_{j=1}^{k_n}|\hat{\alpha}_{n,j}(t)-1|^2 = 0.$$

provided we prove that if either λ_n or μ_n has a limit after translation by some constants A_n, then

(3.17) $$\sup_n \sup_{|t|\leq T}\sum_{j=1}^{k_n}|\hat{\alpha}_{n,j}(t)-1| \leq C < \infty.$$

Let us first suppose that λ_n has a weak limit as $n \to \infty$ after translation by A_n. The characteristic functions

$$\exp\left[\sum_{j=1}^{k_n}(\hat{\alpha}_{n,j}(t)-1) + itA_n\right] = \exp[f_n(t)]$$

have a limit, which is again a characteristic function. Since the limiting characteristic function is continuous and equals 1 at $t = 0$, and the convergence is uniform near 0, on some small interval $|t| \leq T_0$ we have the bound

$$\sup_n \sup_{|t|\leq T_0}[1 - \operatorname{Re} f_n(t)] \leq C$$

or, equivalently,

$$\sup_n \sup_{|t|\leq T_0}\sum_{j=1}^{k_n}\int(1-\cos tx)d\alpha_{n,j} \leq C,$$

and from the subadditivity property $(1 - \cos 2tx) \leq 4(1 - \cos tx)$ this bound extends to an arbitrary interval $|t| \leq T$,

$$\sup_n \sup_{|t|\leq T}\sum_{j=1}^{k_n}\int(1-\cos tx)d\alpha_{n,j} \leq C_T.$$

If we integrate the inequality with respect to t over the interval $[-T, T]$ and divide by $2T$, we get

$$\sup_n \sum_{j=1}^{k_n} \int \left(1 - \frac{\sin Tx}{Tx}\right) d\alpha_{n,j} \leq C_T,$$

from which we can conclude that

$$\sup_n \sum_{j=1}^{k_n} \alpha_{n,j}[|x| \geq \delta] \leq C_\delta < \infty \quad \text{for every } \delta > 0$$

by choosing $T = \frac{2}{\delta}$. Moreover, using the inequality $(1 - \cos x) \geq cx^2$ valid near 0 for a suitable choice of c, we get the estimate

$$\sup_n \sum_{j=1}^{k_n} \int_{|x| \leq 1} x^2 \, d\alpha_{n,j} \leq C < \infty.$$

Now it is straightforward to estimate, for $t \in [-T, T]$,

$$|\hat{\alpha}_{n,j}(t) - 1| = \left| \int [\exp(itx) - 1] d\alpha_{n,j} \right|$$

$$= \left| \int_{|x| \leq 1} [\exp(itx) - 1] d\alpha_{n,j} \right| + \left| \int_{|x| > 1} [\exp(itx) - 1] d\alpha_{n,j} \right|$$

$$\leq \left| \int_{|x| \leq 1} [\exp(itx) - 1 - itx] d\alpha_{n,j} \right|$$

$$+ \left| \int_{|x| > 1} [\exp(itx) - 1] d\alpha_{n,j} \right| + T|a_{n,j}|$$

$$\leq C_1 \int_{|x| \leq 1} x^2 \, d\alpha_{n,j} + C_2 \alpha_{n,j}\left[x : |x| \geq \frac{1}{2}\right],$$

which proves the bound of equation (3.17).

Now we need to establish the same bound under the assumption that μ_n has a limit after suitable translations. For any probability measure α we define $\bar{\alpha}$ by $\bar{\alpha}(A) = \alpha(-A)$ for all Borel sets. The distribution $\alpha * \bar{\alpha}$ is denoted by $|\alpha|^2$. The characteristic functions of $\bar{\alpha}$ and $|\alpha|^2$ are, respectively, $\bar{\hat{\alpha}}(t)$ and $|\hat{\alpha}(t)|^2$ where $\hat{\alpha}(t)$ is the characteristic function of α. An elementary but important fact is $|\alpha * A|^2 = |\alpha|^2$ for any translate A. If μ_n has a limit, so does $|\mu_n|^2$. We conclude that the limit

$$\lim_{n \to \infty} |\hat{\mu}_n(t)|^2 = \lim_{n \to \infty} \prod_{j=1}^{k_n} |\hat{\alpha}_{n,j}(t)|^2$$

exists and defines a characteristic function that is continuous at 0 with a value of 1. Moreover, because of uniform infinitesimality,

$$\lim_{n\to\infty} \inf_{|t|\leq T} |\hat{\alpha}_{n,j}(t)| = 1.$$

It is easy to conclude that there is a $T_0 > 0$ such that

$$\sup_n \sup_{|t|\leq T_0} \sum_{j=1}^{k_n} [1 - |\hat{\alpha}_{n,j}(t)|^2] \leq C_0 < \infty \quad \text{for } |t| \leq T_0$$

and by subadditivity

$$\sup_n \sup_{|t|\leq T} \sum_{j=1}^{k_n} [1 - |\hat{\alpha}_{n,j}(t)|^2] \leq C_T < \infty \quad \text{for any finite } T,$$

providing us with the estimates

(3.18) $$\sup_n \sum_{j=1}^{k_n} |\alpha_{n,j}|^2 [|x| \geq \delta] \leq C_\delta < \infty \quad \text{for any } \delta > 0$$

and

(3.19) $$\sup_n \sum_{j=1}^{k_n} \iint_{|x-y|\leq 2} (x-y)^2 \, d\alpha_{n,j}(x) d\alpha_{n,j}(y) \leq C < \infty.$$

We now show that estimates (3.18) and (3.19) imply (3.17)

$$|\alpha_{n,j}|^2 \left[x : |x| \geq \frac{\delta}{2} \right] \geq \int_{|y|\leq \frac{\delta}{2}} \alpha_{n,j} \left[x : |x-y| \geq \frac{\delta}{2} \right] d\alpha_{n,j}(y)$$

$$\geq \alpha_{n,j}[x : |x| \geq \delta] \alpha_{n,j}\left[x : |x| \leq \frac{\delta}{2} \right]$$

$$\geq \frac{1}{2} \alpha_{n,j}[x : |x| \geq \delta]$$

by uniform infinitesimality. Therefore, (3.18) implies that for every $\delta > 0$,

(3.20) $$\sup_n \sum_{j=1}^{k_n} \alpha_{n,j}[x : |x| \geq \delta] \leq C_\delta < \infty.$$

We now turn to exploiting (3.19). We start with the inequality

$$\iint_{|x-y|\leq 2} (x-y)^2 \, d\alpha_{n,j}(x) d\alpha_{n,j}(y) \geq$$

$$\{\alpha_{n,j}[y : |y| \leq 1]\} \left\{ \inf_{|y|\leq 1} \int_{|x|\leq 1} (x-y)^2 \, d\alpha_{n,j}(x) \right\}.$$

3.8. INFINITELY DIVISIBLE DISTRIBUTIONS

The first term on the right can be assumed to be at least $\frac{1}{2}$ by uniform infinitesimality. The second term

$$\int_{|x|\leq 1} (x-y)^2 \, d\alpha_{n,j}(x) \geq \int_{|x|\leq 1} x^2 \, d\alpha_{n,j}(x) - 2y \int_{|x|\leq 1} x \, d\alpha_{n,j}(x)$$

$$\geq \int_{|x|\leq 1} x^2 \, d\alpha_{n,j}(x) - 2 \left| \int_{|x|\leq 1} x \, d\alpha_{n,j}(x) \right|$$

$$\geq \int_{|x|\leq 1} x^2 \, d\alpha_{n,j}(x) - C\alpha_{n,j}\left[x : |x| \geq \frac{1}{2}\right].$$

The last step is a consequence of estimate (3.16), which we showed we could always assume:

$$\int_{|x|\leq 1} x \, d\alpha_{n,j}(x) \leq C\alpha_{n,j}\left[x : |x| \geq \frac{1}{2}\right].$$

Because of estimate (3.20) we can now assert

(3.21) $$\sup_n \sum_{j=1}^{k_n} \int_{|x|\leq 1} x^2 \, d\alpha_{n,j} \leq C < \infty.$$

One can now derive (3.17) from (3.20) and (3.21) as in the earlier part. □

EXERCISE 3.20. Let $k_n = n^2$ and $\alpha_{n,j} = \delta_{\frac{1}{n}}$ for $1 \leq j \leq n^2$. $\mu_n = \delta_n$ and show that without centering, $\lambda_n * \delta_{-n}$ converges to a different limit.

3.8. Infinitely Divisible Distributions

In the study of limit theorems for sums of independent random variables, infinitely divisible distributions play a very important role.

DEFINITION 3.5 A distribution μ is said to be *infinitely divisible* if for every positive integer n, μ can be written as the *n-fold convolution* $(\lambda_n *)^n$ of some other probability distribution λ_n.

EXERCISE 3.21. Show that the normal distribution with density

$$p(x) = \frac{1}{\sqrt{2\pi}} \exp\left[-\frac{x^2}{2}\right]$$

is infinitely divisible.

EXERCISE 3.22. Show that for any $\lambda \geq 0$, the Poisson distribution with parameter λ

$$p_\lambda(n) = \frac{e^{-n}\lambda^n}{n!} \quad \text{for } n \geq 0$$

is infinitely divisible.

EXERCISE 3.23. Show that any probability distribution supported on a finite set $\{x_1, x_2, \ldots, x_k\}$ with
$$\mu[\{x_j\}] = p_j$$
and $p_j \geq 0$, $\sum_{j=1}^{k} p_j = 1$ is infinitely divisible if and only if it is degenerate, i.e., $\mu[\{x_j\}] = 1$ for some j.

EXERCISE 3.24. Show that for any nonnegative finite measure α with total mass a, the distribution
$$e(F) = e^{-a} \sum_{j=0}^{\infty} \frac{(\alpha *)^j}{j!}$$
with characteristic function
$$\widehat{e(F)}(t) = \exp\left[\int (e^{itx} - 1) d\alpha\right]$$
is an infinitely divisible distribution.

EXERCISE 3.25. Show that the convolution of any two infinitely divisible distributions is again infinitely divisible. In particular, if μ is infinitely divisible, so is any translate $\mu * \delta_a$ for any real a.

We saw in the last section that the asymptotic behavior of $\mu_n * \delta_{A_n}$ can be investigated by means of the asymptotic behavior of $\lambda_n * \delta_{A_n}$, and the characteristic function $\hat{\lambda}_n$ of λ_n has a very special form,

$$\hat{\lambda}_n = \prod_{j=1}^{k_n} \exp[\hat{\beta}_{n,j}(t) - 1 + ita_{n,j}]$$

$$= \exp\left[\sum_{j=1}^{k_n} \int [e^{itx} - 1] d\beta_{n,j} + it \sum_{j=1}^{k_n} a_{n,j}\right]$$

$$= \exp\left[\int [e^{itx} - 1] dM_n + ita_n\right]$$

$$= \exp\left[\int [e^{itx} - 1 - it\theta(x)] dM_n + it\left[\int \theta(x) dM_n + a_n\right]\right]$$

(3.22)
$$= \exp\left[\int [e^{itx} - 1 - it\theta(x)] dM_n + itb_n\right].$$

We can make any reasonable choice for $\theta(\cdot)$, and we will need it to be a bounded continuous function with
$$|\theta(x) - x| \leq C|x|^3$$
near 0. Possible choices are $\theta(x) = \frac{x}{1+x^2}$, or $\theta(x) = x$ for $|x| \leq 1$ and $\text{sign}(x)$ for $|x| \geq 1$. We now investigate when such things will have a weak limit. Convoluting with δ_{A_n} only changes b_n to $b_n + A_n$.

3.8. INFINITELY DIVISIBLE DISTRIBUTIONS

First we note that

$$\text{\Large ※} \quad \hat{\mu}(t) = \exp\left[\int [e^{itx} - 1 - it\theta(x)]dM + ita\right]$$

is a characteristic function for any measure M with finite total mass. In fact, it is the characteristic function of an infinitely divisible probability distribution. It is not necessary that M be a finite measure for μ to make sense. M could be infinite, but in such a way that it is finite on $\{x : |x| \geq \delta\}$ for every $\delta > 0$, and near 0 it integrates x^2, i.e.,

(3.23) $$M[x : |x| \geq \delta] < \infty \quad \text{for all } \delta > 0,$$

(3.24) $$\int_{|x| \leq 1} x^2 \, dM < \infty.$$

To see this we remark that

$$\hat{\mu}_\delta(t) = \exp\left[\int_{|x| \geq \delta} [e^{itx} - 1 - it\theta(x)]dM + ita\right]$$

is a characteristic function for each $\delta > 0$ and because

$$|e^{itx} - 1 - itx| \leq C_T x^2$$

for $|t| \leq T$, $\hat{\mu}_\delta(t) \to \hat{\mu}(t)$ uniformly on bounded intervals where $\hat{\mu}(t)$ is given by the integral

$$\hat{\mu}(t) = \exp\left[\int [e^{itx} - 1 - it\theta(x)]dM + ita\right],$$

which converges absolutely and defines a characteristic function. Let us call measures that satisfy (3.23) and (3.24) which can be expressed in the form

(3.25) $$\int \frac{x^2}{1+x^2} dM < \infty$$

admissible *Lévy measures*. Since the same argument applies to $\frac{M}{n}$ and $\frac{a}{n}$ instead of M and a, for any admissible Lévy measure M and real number a, $\hat{\mu}(t)$ is in fact, an infinitely divisible characteristic function. As the normal distribution is also an infinitely divisible probability distribution, we arrive at the following:

THEOREM 3.20 *For every admissible Lévy measure M, $\sigma^2 > 0$, and real a,*

$$\hat{\mu}(t) = \exp\left[\int [e^{itx} - 1 - it\theta(x)]dM + ita - \frac{\sigma^2 t^2}{2}\right]$$

is the characteristic function of an infinitely divisible distribution μ.

We will denote this distribution μ by $\mu = \mathbf{e}(M, \sigma^2, a)$. The main theorem of this section is

62   3. INDEPENDENT SUMS

THEOREM 3.21 *In order that $\mu_n = \mathbf{e}(M_n, \sigma_n^2, a_n)$ may converge to a limit μ, it is necessary and sufficient that $\mu = \mathbf{e}(M, \sigma^2, a)$ and the following three conditions (3.26), (3.27), and (3.28) are satisfied.*

For every bounded continuous function f that vanishes in some neighborhood of 0,

$$(3.26) \qquad \lim_{n\to\infty} \int f(x)dM_n = \int f(x)dM.$$

For some (and therefore for every) $\ell > 0$ such that $\pm\ell$ are continuity points for M, i.e., $M\{\pm\ell\} = 0$,

$$(3.27) \qquad \lim_{n\to\infty}\left[\sigma_n^2 + \int_{-\ell}^{\ell} x^2\, dM_n\right] = \left[\sigma^2 + \int_{-\ell}^{\ell} x^2\, dM\right]$$

$$(3.28) \qquad a_n \to a \quad \text{as } n \to \infty.$$

PROOF: Let us prove the sufficiency first. Condition (3.26) implies that for every ℓ such that $\pm\ell$ are continuity points of M,

$$\lim_{n\to\infty}\int_{|x|\ge\ell} [e^{itx} - 1 - it\theta(x)]dM_n = \int_{|x|\ge\ell} [e^{itx} - 1 - it\theta(x)]dM,$$

and because of condition (3.27), it is enough to show that

$$\lim_{\ell\to 0}\limsup_{n\to\infty}\left|\int_{-\ell}^{\ell}\left[e^{itx} - 1 - it\theta(x) + \frac{t^2x^2}{2}\right]dM_n \right.$$
$$\left. - \int_{-\ell}^{\ell}\left[e^{itx} - 1 - it\theta(x) + \frac{t^2x^2}{2}\right]dM\right| = 0$$

in order to conclude that

$$\lim_{n\to\infty}\left[-\frac{\sigma_n^2 t^2}{2} + \int [e^{itx} - 1 - it\theta(x)]dM_n\right] =$$
$$\left[-\frac{\sigma^2 t^2}{2} + \int [e^{itx} - 1 - it\theta(x)]dM\right].$$

This follows from the estimates

$$\left|e^{itx} - 1 - it\theta(x) + \frac{t^2x^2}{2}\right| \le C_T|x|^3 \quad \text{and} \quad \int_{-\ell}^{\ell}|x|^3 dM_n \le \ell\int_{-\ell}^{\ell}|x|^2 dM_n.$$

Condition (3.28) takes care of the terms involving a_n.

We now turn to proving the necessity. If μ_n has a weak limit μ, then the absolute values of the characteristic functions $|\hat{\mu}_n(t)|$ are all uniformly close to 1 near 0. Since

$$|\hat{\mu}_n(t)| = \exp\left[-\int(1 - \cos tx)dM_n - \frac{\sigma_n^2 t^2}{2}\right],$$

taking logarithms we conclude that

$$\limsup_{t\to 0}{}_n\left[\frac{\sigma_n t^2}{2} + \int(1 - \cos tx)dM_n\right] = 0.$$

3.8. INFINITELY DIVISIBLE DISTRIBUTIONS

This implies (3.29), (3.30), and (3.31). For each $\ell > 0$,

(3.29) $$\sup_n M_n\{x : |x| \geq \ell\} < \infty,$$

(3.30) $$\lim_{A \to \infty} \sup_n M_n\{x : |x| \geq A\} = 0.$$

For every $0 \leq \ell < \infty$,

(3.31) $$\sup_n \left[\sigma_n^2 + \int_{-\ell}^{\ell} |x|^2 dM_n\right] < \infty.$$

We can choose a subsequence of M_n (which we will denote by M_n as well) that "converges" in the sense that it satisfies conditions (3.26) and (3.27) of the theorem. Then $\mathbf{e}(M_n, \sigma_n^2, 0)$ converges weakly to $\mathbf{e}(M, \sigma^2, 0)$. It is not hard to see that for any sequence of probability distributions α_n if both α_n and $\alpha_n * \delta_{a_n}$ converge to limits α and β, respectively, then, necessarily, $\beta = \alpha * \delta_a$ for some a and $a_n \to a$ as $n \to \infty$. In order complete the proof of necessity, we need only establish the uniqueness of the representation, which is done in the next lemma. □

LEMMA 3.22 (Uniqueness) *Suppose* $\mu = \mathbf{e}(M_1, \sigma_1^2, a_1) = \mathbf{e}(M_2, \sigma_2^2, a_2)$; *then* $M_1 = M_2$, $\sigma_1^2 = \sigma_2^2$, *and* $a_1 = a_2$.

PROOF: Since $\hat{\mu}(t)$ never vanishes, by taking logarithms we have

(3.32) $$\psi(t) = \left[-\frac{\sigma_1^2 t^2}{2} + \int [e^{itx} - 1 - it\theta(x)]dM_1 + ita_1\right]$$
$$= \left[-\frac{\sigma_2^2 t^2}{2} + \int [e^{itx} - 1 - it\theta(x)]dM_2 + ita_2\right].$$

We can verify that for any admissible Lévy measure M

$$\lim_{t \to \infty} \frac{1}{t^2} \int [e^{itx} - 1 - it\theta(x)]dM = 0.$$

Consequently,

$$\lim_{t \to \infty} \frac{\psi(t)}{t^2} = \sigma_1^2 = \sigma_2^2$$

leaving us with

$$\psi(t) = \left[\int [e^{itx} - 1 - it\theta(x)]dM_1 + ita_1\right]$$
$$= \left[\int [e^{itx} - 1 - it\theta(x)]dM_2 + ita_2\right]$$

for a different ψ. If we calculate

$$H(s, t) = \frac{\psi(t+s) + \psi(t-s)}{2} - \psi(t),$$

we get

$$\int e^{itx}(1 - \cos sx)dM_1 = \int e^{itx}(1 - \cos sx)dM_2$$

for all t and s. Since we can and do assume that $M\{0\} = 0$ for any admissible Lévy measure M, we have $M_1 = M_2$. If we know that $\sigma_1^2 = \sigma_2^2$ and $M_1 = M_2$, it is easy to see that a_1 must equal a_2. □

Finally, we have the following:

COROLLARY 3.23 (Lévy-Khintchine Representation) *Any infinitely divisible distribution admits a representation* $\mu = \mathbf{e}(M, \sigma^2, a)$ *for some admissible Lévy measure* M, $\sigma^2 > 0$, *and real number* a.

PROOF: We can write $\mu = \mu_n *^n = \mu_n * \mu_n * \cdots * \mu_n$ with n terms. If we show that $\mu_n \Rightarrow \delta_0$, then the sequence is uniformly infinitesimal and by the earlier theorem on accompanying laws μ will be the limit of some $\lambda_n = \mathbf{e}(M_n, 0, a_n)$ and therefore has to be of the form $\mathbf{e}(M, \sigma^2, a)$ for some choice of admissible Lévy measure M, $\sigma^2 > 0$, and real a. In a neighborhood around 0, $\hat{\mu}(t)$ is close to 1 and it is easy to check that

$$\hat{\mu}_n(t) = [\hat{\mu}(t)]^{1/n} \to 1$$

as $n \to \infty$ in that neighborhood. As we saw before, this implies that $\mu_n \Rightarrow \delta_0$. □

Applications.

(1) Convergence to the Poisson distribution. Let $\{X_{n,j} : 1 \leq j \leq k_n\}$ be k_n independent random variables taking the values 0 or 1 with probabilities $1 - p_{n,j}$ and $p_{n,j}$, respectively. We assume that

$$\lim_{n \to \infty} \sup_{1 \leq j \leq k_n} p_{n,j} = 0,$$

which is the uniform infinitesimality condition. We are interested in the limiting distribution of $S_n = \sum_{j=1}^{k_n} X_{n,j}$ as $n \to \infty$. Since we have to center by the mean, we can pick any level, say $\frac{1}{2}$, for truncation; then the truncated means are all 0. The accompanying laws are given by $\mathbf{e}(M_n, 0, a_n)$ with $M_n = (\sum p_{n,j})\delta_1$ and $a_n = (\sum p_{n,j})\theta(1)$. It is clear that a limit exists if and only if $\lambda_n = \sum p_{n,j}$ has a limit λ as $n \to \infty$ and the limit in such a case is the Poisson distribution with parameter λ.

(2) Convergence to the normal distribution. If the limit of $S_n = \sum_{j=1}^{k_n} X_{n,j}$ of k_n uniformly infinitesimal mutually independent random variables exists, then the limit is normal if and only if $M \equiv 0$. If $a_{n,j}$ is the centering needed, this is equivalent to

$$\lim_{n \to \infty} \sum_j P[|X_{n,j} - a_{n,j}| \geq \varepsilon] = 0 \quad \text{for all } \varepsilon > 0.$$

Since $\lim_{n \to \infty} \sup_j |a_{n,j}| = 0$, this is equivalent to

$$\lim_{n \to \infty} \sum_j P[|X_{n,j}| \geq \varepsilon] = 0 \quad \text{for each } \varepsilon > 0.$$

(3) The limiting variance and the mean are given by

$$\sigma^2 = \lim_{n\to\infty} \sum_j E\{[X_{n,j} - a_{n,j}]^2 : |X_{n,j} - a_{n,j}| \le 1\} \quad \text{and} \quad a = \lim_{n\to\infty} \sum_j a_{n,j}$$

where

$$a_{n,j} = \int_{|x|\le 1} x\, d\alpha_{n,j}.$$

Suppose that $E[X_{n,j}] = 0$ for all $1 \le j \le k_n$ and n. Assume that $\sigma_n^2 = \sum_j E\{[X_{n,j}]^2\}$ and $\sigma^2 = \lim_{n\to\infty} \sigma_n^2$ exists. What do we need in order to make sure that the limiting distribution is normal with mean 0 and variance σ^2? Let $\alpha_{n,j}$ be the distribution of $X_{n,j}$.

$$|a_{n,j}|^2 = \left|\int_{|x|\le 1} x\, d\alpha_{n,j}\right|^2 = \left|\int_{|x|>1} x\, d\alpha_{n,j}\right|^2 \le \alpha_{n,j}[|x|>1] \int |x|^2\, d\alpha_{n,j}$$

and

$$\sum_{j=1}^{k_n} |a_{n,j}|^2 \le \left\{\sum_{1\le j\le k_n} \int |x|^2\, d\alpha_{n,j}\right\}\left\{\sup_{1\le j\le k_n} \alpha_{n,j}[|x|>1]\right\}$$

$$\le \sigma_n^2 \left\{\sup_{1\le j\le k_n} \alpha_{n,j}[|x|>1]\right\} \to 0.$$

Because $\sum_{j=1}^{k_n} |a_{n,j}|^2 \to 0$ as $n \to \infty$, we must have

$$\sigma^2 = \lim_{n\to\infty} \sum_j \int_{|x|\le\ell} |x|^2\, d\alpha_{n,j} \quad \text{for every } \ell > 0,$$

or, equivalently,

$$\lim_{n\to\infty} \sum_j \int_{|x|>\ell} |x|^2\, d\alpha_{n,j} = 0 \quad \text{for every } \ell$$

establishing the necessity as well as sufficiency in Lindeberg's theorem. A simple calculation shows that

$$\sum_j |a_{n,j}| \le \sum_j \int_{|x|>1} |x|\, d\alpha_{n,j} \le \sum_j \int_{|x|>1} |x|^2\, d\alpha_{n,j} = 0$$

establishing that the limiting normal distribution has mean 0.

EXERCISE 3.26. What happens in the Poisson limit theorem (application (1)) if $\lambda_n = \sum_j p_{n,j} \to \infty$ as $n \to \infty$? Can you show that the distribution of $\frac{S_n - \lambda_n}{\sqrt{\lambda_n}}$ converges to the standard normal distribution?

3.9. Laws of the Iterated Logarithm

When we are dealing with a sequence of independent identically distributed random variables $X_1, X_2, \ldots, X_n, \ldots$ with mean 0 and variance 1, we have a strong law of large numbers asserting that

$$P\left\{\lim_{n \to \infty} \frac{X_1 + X_2 + \cdots + X_n}{n} = 0\right\} = 1$$

and a central limit theorem asserting that

$$P\left\{\frac{X_1 + X_2 + \cdots + X_n}{\sqrt{n}} \leq a\right\} \to \int_{-\infty}^{a} \frac{1}{\sqrt{2\pi}} \exp\left[-\frac{x^2}{2}\right] dx .$$

It is a reasonable question to ask if the random variables $\frac{X_1+X_2+\cdots+X_n}{\sqrt{n}}$ themselves converge to some limiting random variable Y that is distributed according to the standard normal distribution. The answer is no and is not hard to show.

LEMMA 3.24 *For any sequence n_j of numbers $\to \infty$,*

$$P\left\{\limsup_{j \to \infty} \frac{X_1 + X_2 + \cdots + X_{n_j}}{\sqrt{n_j}} = +\infty\right\} = 1$$

PROOF: Let us define

$$Z = \limsup_{j \to \infty} \frac{X_1 + X_2 + \cdots + X_{n_j}}{\sqrt{n_j}}$$

which can be $+\infty$. Because the normal distribution has an infinitely long tail, i.e., the probability of exceeding any given value is positive, we must have

$$P[Z \geq a] > 0$$

for any a. But Z is a random variable that does not depend on the particular values of $X_1, X_2, \ldots, X_k$ and therefore $[Z \geq a]$ is a set in the tail σ-field. By Kolmogorov's zero-one law $P[Z \geq a]$ must be either 0 or 1. Since it cannot be 0 it must be 1. □

Since we know that $\frac{X_1+X_2+\cdots+X_n}{n} \to 0$ with probability 1 as $n \to \infty$, the question arises as to the rate at which this happens. The law of the iterated logarithm provides an answer.

THEOREM 3.25 *For any sequence $X_1, X_2, \ldots, X_n, \ldots$ of independent identically distributed random variables with mean 0 and variance 1,*

$$P\left\{\limsup_{n \to \infty} \frac{X_1 + X_2 + \cdots + X_n}{\sqrt{n \log \log n}} = \sqrt{2}\right\} = 1 .$$

We will not prove this theorem in the most general case, which assumes only the existence of two moments. We will assume instead that $E[|X|^{2+\alpha}] < \infty$ for some $\alpha > 0$. We shall first reduce the proof to an estimate on the tail behavior of the distributions of $\frac{S_n}{\sqrt{n}}$ by a careful application of the Borel-Cantelli lemma. This estimate is obvious if $X_1, X_2, \ldots, X_n, \ldots$ are themselves normally distributed, and

3.9. LAWS OF THE ITERATED LOGARITHM

we will show how to extend it to a large class of distributions that satisfy the additional moment condition. It is clear that we are interested in showing that

$$P\{S_n \geq \lambda\sqrt{n \log \log n} \text{ infinitely often}\} = 0 \quad \text{for } \lambda > \sqrt{2}.$$

It would be sufficient because of the Borel-Cantelli lemma to show that

$$\sum_n P\{S_n \geq \lambda\sqrt{n \log \log n}\} < \infty \quad \text{for any } \lambda > \sqrt{2}.$$

This, however, is too strong. The condition of the Borel-Cantelli lemma is not necessary in this context because of the strong dependence between the partial sums S_n. The function $\phi(n) = \sqrt{n \log \log n}$ is clearly well-defined and nondecreasing for $n \geq 3$, and it is sufficient for our purposes to show that for any $\lambda > \sqrt{2}$ we can find some sequence $k_n \uparrow \infty$ of integers such that

(3.33) $$\sum_n P\left\{\sup_{k_{n-1} \leq j \leq k_n} S_j \geq \lambda \phi(k_{n-1})\right\} < \infty.$$

This will establish that with probability 1,

$$\limsup_{n \to \infty} \frac{\sup_{k_{n-1} \leq j \leq k_n} S_j}{\phi(k_{n-1})} \leq \lambda$$

or, by the monotonicity of ϕ,

$$\limsup_{n \to \infty} \frac{S_n}{\phi(n)} \leq \lambda$$

with probability 1. Since $\lambda > \sqrt{2}$ is arbitrary, the upper bound in the law of the iterated logarithm will follow. Each term in the sum of (3.33) can be estimated as in Lévy's inequality,

$$P\left\{\sup_{k_{n-1} \leq j \leq k_n} S_j \geq \lambda \phi(k_{n-1})\right\} \leq 2P\{S_{k_n} \geq (\lambda - \sigma)\phi(k_{n-1})\}$$

with $0 < \sigma < \lambda$, provided

$$\sup_{1 \leq j \leq k_n - k_{n-1}} P\{|S_j| \geq \sigma \phi(k_{n-1})\} \leq \frac{1}{2}.$$

Our choice of k_n will be $k_n = [\rho^n]$ for some $\rho > 1$ and therefore

$$\lim_{n \to \infty} \frac{\phi(k_{n-1})}{\sqrt{k_n}} = \infty,$$

and, by Chebyshev's inequality, for any fixed $\sigma > 0$,

$$\sup_{1 \leq j \leq k_n} P\{|S_j| \geq \sigma \phi(k_{n-1})\} \leq \frac{E[S_n^2]}{[\sigma \phi(k_{n-1})]^2} = \frac{k_n}{[\sigma \phi(k_{n-1})]^2}$$

$$= \frac{k_n}{\sigma^2 k_{n-1} \log \log k_{n-1}}$$

(3.34) $$= o(1) \quad \text{as } n \to \infty.$$

By choosing σ small enough so that $\lambda - \sigma > \sqrt{2}$, it is sufficient to show that

$$\sum_n P\{S_{k_n} \geq \lambda'\phi(k_{n-1})\} < \infty \quad \text{for any } \lambda' > \sqrt{2}.$$

By picking ρ sufficiently close to 1 (so that $\lambda'\sqrt{\rho} > \sqrt{2}$), because $\frac{\phi(k_{n-1})}{\phi(k_n)} = \frac{1}{\sqrt{\rho}}$ we can reduce this to the convergence of

(3.35) $$\sum_n P\{S_{k_n} \geq \lambda\phi(k_n)\} < \infty \quad \text{for all } \lambda > \sqrt{2}.$$

If we use the estimate $P[X \geq a] \leq \exp[-\frac{a^2}{2}]$ that is valid for the standard normal distribution, we can verify (3.35).

$$\sum_n \exp\left[-\frac{\lambda^2(\phi(k_n))^2}{2k_n}\right] < \infty \quad \text{for any } \lambda > \sqrt{2}.$$

To prove the lower bound we select again a subsequence, $k_n = [\rho^n]$ with some $\rho > 1$, and look at $Y_n = S_{k_{n+1}} - S_{k_n}$, which are now independent random variables. The tail probability of the normal distribution has the lower bound

$$P[X \geq a] = \frac{1}{\sqrt{2\pi}} \int_a^\infty \exp\left[-\frac{x^2}{2}\right] dx$$
$$\geq \frac{1}{\sqrt{2\pi}} \int_a^\infty \exp\left[-\frac{x^2}{2} - x\right](x+1)dx \geq \frac{1}{\sqrt{2\pi}} \exp\left[-\frac{(a+1)^2}{2}\right].$$

If we assume normal-like tail probabilities, we can conclude that

$$\sum_n P\{Y_n \geq \lambda\phi(k_{n+1})\} \geq \sum_n \exp\left[-\frac{1}{2}\left[1 + \frac{\lambda\phi(k_{n+1})}{\sqrt{(\rho^{n+1} - \rho^n)}}\right]^2\right] = +\infty$$

provided $\frac{\lambda^2\rho}{2(\rho-1)} < 1$ and conclude by the Borel-Cantelli lemma that $Y_n = S_{k_{n+1}} - S_{k_n}$ exceeds $\lambda\phi(k_{n+1})$ infinitely often for such λ. On the other hand, from the upper bound we already have (replacing X_i by $-X_i$)

$$P\left\{\limsup_n \frac{-S_{k_n}}{\phi(k_{n+1})} \leq \frac{\sqrt{2}}{\sqrt{\rho}}\right\} = 1.$$

Consequently,

$$P\left\{\limsup_n \frac{S_{k_{n+1}}}{\phi(k_{n+1})} \geq \sqrt{\frac{2(\rho-1)}{\rho}} - \frac{\sqrt{2}}{\sqrt{\rho}}\right\} = 1,$$

and therefore

$$P\left\{\limsup_n \frac{S_n}{\phi(n)} \geq \sqrt{\frac{2(\rho-1)}{\rho}} - \frac{\sqrt{2}}{\sqrt{\rho}}\right\} = 1.$$

We now take ρ arbitrarily large and we are done.

We saw that the law of the iterated logarithm depends on two things:

3.9. LAWS OF THE ITERATED LOGARITHM

(1) For any $a > 0$ and $p < \frac{a^2}{2}$ an upper bound for the probability,

$$P[S_n \geq a\sqrt{n \log \log n}] \leq C_p [\log n]^{-p}$$

with some constant C_p, and

(2) for any $a > 0$ and $p > \frac{a^2}{2}$ a lower bound for the probability,

$$P[S_n \geq a\sqrt{n \log \log n}] \geq C_p [\log n]^{-p}$$

with some, possibly different, constant C_p.

Both inequalities can be obtained from a uniform rate of convergence in the central limit theorem.

$$(3.36) \qquad \sup_a \left| P\left\{ \frac{S_n}{\sqrt{n}} \geq a \right\} - \int_a^\infty \frac{1}{\sqrt{2\pi}} \exp\left[-\frac{x^2}{2}\right] dx \right| \leq C n^{-\delta}$$

for some $\delta > 0$ in the central limit theorem. Such an error estimate is provided in the following theorem:

THEOREM 3.26 (Berry-Esseen Theorem) *Assume that the i.i.d. sequence $\{X_j\}$ with mean 0 and variance 1 satisfies an additional moment condition $E|X|^{2+\alpha} < \infty$ for some $\alpha > 0$. Then for some $\delta > 0$ the estimate (3.36) holds.*

PROOF: The proof will be carried out after two lemmas.

LEMMA 3.27 *Let $-\infty < a < b < \infty$ be given and $0 < h < \frac{b-a}{2}$ be a small positive number. Consider the function $f_{a,b,h}(x)$ defined as*

$$f_{a,b,h}(x) = \begin{cases} 0 & \text{for } -\infty < x \leq a - h \\ \frac{x-a+h}{2h} & \text{for } a - h \leq x \leq a + h \\ 1 & \text{for } a + h \leq x \leq b - h \\ 1 - \frac{x-b+h}{2h} & \text{for } b - h \leq x \leq b + h \\ 0 & \text{for } b + h \leq x < \infty. \end{cases}$$

For any probability distribution μ with characteristic function $\hat{\mu}(t)$

$$\int_{-\infty}^\infty f_{a,b,h}(x) d\mu(x) = \frac{1}{2\pi} \int_{-\infty}^\infty \hat{\mu}(y) \frac{e^{-iay} - e^{-iby}}{iy} \frac{\sin hy}{hy} dy.$$

PROOF: This is essentially the Fourier inversion formula. Note that

$$f_{a,b,h}(x) = \frac{1}{2\pi} \int_{-\infty}^\infty e^{ixy} \frac{e^{-iay} - e^{-iby}}{iy} \frac{\sin hy}{hy} dy.$$

We can start with the double integral

$$\frac{1}{2\pi} \int_{-\infty}^\infty \int_{-\infty}^\infty e^{ixy} \frac{e^{-iay} - e^{-iby}}{iy} \frac{\sin hy}{hy} dy \, d\mu(x)$$

and apply Fubini's theorem to obtain the lemma. □

LEMMA 3.28 *If λ and μ are two probability measures with zero mean having $\hat{\lambda}(\cdot)$ and $\hat{\mu}(\cdot)$ for respective characteristic functions, then*

$$\int_{-\infty}^{\infty} f_{a,h}(x) d(\lambda - \mu)(x) = \frac{1}{2\pi} \int_{-\infty}^{\infty} [\hat{\lambda}(y) - \hat{\mu}(y)] \frac{e^{-iay}}{iy} \frac{\sin hy}{hy} dy$$

where $f_{a,h}(x) = f_{a,\infty,h}(x)$ is given by

$$f_{a,h}(x) = \begin{cases} 0 & \text{for } -\infty < x \le a-h \\ \frac{x-a+h}{2h} & \text{for } a-h \le x \le a+h \\ 1 & \text{for } a+h \le x < \infty. \end{cases}$$

PROOF: We just let $b \to \infty$ in the previous lemma. Since $|\hat{\lambda}(y) - \hat{\mu}(y)| = o(|y|)$, there is no problem in applying the Riemann-Lebesgue lemma.

We now proceed with the proof of the theorem.

$$\lambda[[a,\infty)] \le \int f_{a-h,h}(x) d\lambda(x) \le \lambda[[a-2h,\infty)]$$

and

$$\mu[[a,\infty)] \le \int f_{a-h,h}(x) d\mu(x) \le \mu[[a-2h,\infty)].$$

Therefore, if we assume that μ has a density bounded by C,

$$\lambda[[a,\infty)] - \mu[[a,\infty)] \le 2hC + \int f_{a-h,h}(x) d(\lambda - \mu)(x).$$

Since we get a similar bound in the other direction as well,

$$\sup_a |\lambda[[a,\infty)] - \mu[[a,\infty)]|$$

(3.37)
$$\le \sup_a \left| \int f_{a-h,h}(x) d(\lambda - \mu)(x) \right| + 2hC$$

$$\le \frac{1}{2\pi} \int_{-\infty}^{\infty} |\hat{\lambda}(y) - \hat{\mu}(y)| \frac{|\sin hy|}{hy^2} dy + 2hC.$$

□

Now we return to the proof of the theorem. We take λ to be the distribution of $\frac{S_n}{\sqrt{n}}$ having as its characteristic function $\hat{\lambda}_n(y) = [\phi(\frac{y}{\sqrt{n}})]^n$ where $\phi(y)$ is the characteristic function of the common distribution of the $\{X_i\}$ and has the expansion

$$\phi(y) = 1 - \frac{y^2}{2} + O(|y|^{2+\alpha})$$

for some $\alpha > 0$. We therefore get, for some choice of $\alpha > 0$,

$$\left| \hat{\lambda}_n(y) - \exp\left[-\frac{y^2}{2}\right] \right| \le C \frac{|y|^{2+\alpha}}{n^\alpha} \quad \text{if } |y| \le n^{\frac{\alpha}{2+\alpha}}.$$

Therefore, for $\theta = \frac{\alpha}{2+\alpha}$

$$\int_{-\infty}^{\infty} \left|\hat{\lambda}_n(y) - \exp\left[-\frac{y^2}{2}\right]\right| \frac{|\sin hy|}{hy^2} dy$$

$$= \int_{|y| \leq n^\theta} \left|\hat{\lambda}_n(y) - \exp\left[-\frac{y^2}{2}\right]\right| \frac{|\sin hy|}{hy^2} dy$$

$$+ \int_{|y| \geq n^\theta} \left|\hat{\lambda}_n(y) - \exp\left[-\frac{y^2}{2}\right]\right| \frac{|\sin hy|}{hy^2} dy$$

$$\leq \frac{C}{h} \left\{ \int_{|y| \leq n^\theta} \frac{|y|^\alpha}{n^\alpha} dy + \int_{|y| \geq n^\theta} \frac{dy}{|y|^2} \right\}$$

$$\leq C \frac{n^{(\alpha+1)\theta - \alpha} + n^{-\theta}}{h} = \frac{C}{hn^{\frac{\alpha}{\alpha+2}}}.$$

Substituting this bound in (3.37) we get

$$\sup_a |\lambda_n[[a, \infty)] - \mu[[a, \infty)]| \leq C_1 h + \frac{C}{hn^{\frac{\alpha}{2+\alpha}}}.$$

By picking $h = n^{-\frac{\alpha}{2(2+\alpha)}}$ we get

$$\sup_a |\lambda_n[[a, \infty)] - \mu[[a, \infty)]| \leq Cn^{-\frac{\alpha}{2(2+\alpha)}}$$

and we are done. $\square$

CHAPTER 4

Dependent Random Variables

4.1. Conditioning

One of the key concepts in probability theory is the notion of conditional probability and conditional expectation. Suppose that we have a probability space $(\Omega, \mathcal{F}, P)$ consisting of a space Ω, a σ-field $\mathcal{F}$ of subsets of Ω, and a probability measure on the σ-field $\mathcal{F}$. If we have a set $A \in \mathcal{F}$ of positive measure, then conditioning with respect to A means we restrict ourselves to the set A. Ω gets replaced by A and the σ-field $\mathcal{F}$ by the σ-field $\mathcal{F}_A$ of subsets of A that are in $\mathcal{F}$. For $B \subset A$ we define

$$P_A(B) = \frac{P(B)}{P(A)}.$$

We could achieve the same thing by defining for arbitrary $B \in \mathcal{F}$

$$P_A(B) = \frac{P(A \cap B)}{P(A)},$$

in which case $P_A(\cdot)$ is a measure defined on $\mathcal{F}$ as well but one that is concentrated on A and assigning 0 probability to A^c. The definition of conditional probability is

$$P(B \mid A) = \frac{P(A \cap B)}{P(A)}.$$

Similarly, the definition of conditional expectation of an integrable function $f(\omega)$ given a set $A \in \mathcal{F}$ of positive probability is defined to be

$$E\{f \mid A\} = \frac{\int_A f(\omega) dP}{P(A)}.$$

In particular, if we take $f = \chi_B$ for some $B \in \mathcal{F}$ we recover the definition of conditional probability. In general, if we know $P(B \mid A)$ and $P(A)$, we can recover $P(A \cap B) = P(A)P(B \mid A)$ but we cannot recover $P(B)$. But if we know $P(B \mid A)$ as well as $P(B \mid A^c)$ along with $P(A)$ and $P(A^c) = 1 - P(A)$, then

$$P(B) = P(A \cap B) + P(A^c \cap B) = P(A)P(B \mid A) + P(A^c)P(B \mid A^c).$$

More generally, if $\mathcal{P}$ is a partition of Ω into a finite or even a countable number of disjoint measurable sets $A_1, A_2, \ldots, A_j, \ldots,$

$$P(B) = \sum_j P(A_j) P(B \mid A_j).$$

If ξ is a random variable taking distinct values $\{a_j\}$ on $\{A_j\}$, then

$$P(B \mid \{\xi = a_j\}) = P(B \mid A_j),$$

or, more generally,

$$* \quad P(B \mid \{\xi = a\}) = \frac{P(B \cap \{\xi = a\})}{P(\xi = a)},$$

provided $P(\xi = a) > 0$. One of our goals is to seek a definition that makes sense when $P(\xi = a) = 0$. This involves dividing 0 by 0 and should involve differentiation of some kind. In the countable case we may think of $P(B \mid \xi = a_j)$ as a function $f_B(\xi)$ that is equal to $P(B \mid A_j)$ on $\xi = a_j$. We can rewrite our definition of

$$f_B(a_j) = P(B \mid \{\xi = a_j\})$$

as

$$\int_{\xi = a_j} f_B(\xi) dP = P(B \cap \{\xi = a_j\}) \quad \text{for each } j$$

or, summing over any arbitrary collection of j's,

$$\int_{\xi \in E} f_B(\xi) dP = P(B \cap \{\xi \in E\}).$$

Sets of the form $\xi \in E$ form a sub σ-field $\Sigma \subset \mathcal{F}$, and we can rewrite the definition as

$$\int_A f_B(\xi) dP = P(B \cap A) \quad \text{for all } A \in \Sigma.$$

Of course, in this case $A \in \Sigma$ if and only if A is a union of the atoms $\xi = a$ of the partition over a finite or countable subcollection of the possible values of a. Similar considerations apply to the conditional expectation of a random variable G given ξ. The equation becomes

$$\int_A g(\xi) dP = \int_A G(\omega) dP,$$

or we can rewrite this as

$$\int_A g(\omega) dP = \int_A G(\omega) dP$$

for all $A \in \Sigma$, and instead of demanding that g be a function of ξ, we demand that g be Σ-measurable, which is the same thing. Now the random variable ξ is out of the picture, and rightly so. What is important is the information we have if we know ξ, and that is the same if we replace ξ by a one-to-one function of itself. The σ-field Σ abstracts that information nicely. So it turns out that the proper notion of conditioning involves a sub σ-field $\Sigma \subset \mathcal{F}$. If G is an integrable function and $\Sigma \subset \mathcal{F}$ is given, we will seek another integrable function g that is Σ-measurable and satisfies

$$\int_A g(\omega) dP = \int_A G(\omega) dP$$

for all $A \in \Sigma$. We will prove existence and uniqueness of such a g and call it the conditional expectation of G given Σ and denote it by $g = E[G \mid \Sigma]$.

The way to prove the above result will take us on a detour. A signed measure on a measurable space $(\Omega, \mathcal{F})$ is a set function $\lambda(\cdot)$ defined for $A \in \mathcal{F}$ that is countably additive but not necessarily nonnegative. Countable additivity is again in any of the following two equivalent senses:

$$\lambda\left(\bigcup A_n\right) = \sum \lambda(A_n)$$

for any countable collection of disjoint sets in $\mathcal{F}$, or

$$\lim_{n \to \infty} \lambda(A_n) = \lambda(A)$$

whenever $A_n \downarrow A$ or $A_n \uparrow A$.

Examples of such λ can be constructed by taking the difference $\mu_1 - \mu_2$ of two nonnegative measures μ_1 and μ_2.

DEFINITION 4.1 A set $A \in \mathcal{F}$ is *totally positive* (*totally negative*) for λ if for every subset $B \in \mathcal{F}$ with $B \subset A$, $\lambda(B) \geq 0$ (≤ 0).

REMARK 4.1. A measurable subset of a totally positive set is totally positive. Any countable union of totally positive subsets is again totally positive.

LEMMA 4.1 *If λ is a countably additive signed measure on $(\Omega, \mathcal{F})$,*

$$\sup_{A \in \mathcal{F}} |\lambda(A)| < \infty.$$

PROOF: The key idea in the proof is that, since $\lambda(\Omega)$ is a finite number, if $\lambda(A)$ is large, so is $\lambda(A^c)$ with an opposite sign. In fact, it is not hard to see that $||\lambda(A)| - |\lambda(A^c)|| \leq |\lambda(\Omega)|$ for all $A \in \mathcal{F}$. Another fact is that if $\sup_{B \subset A} |\lambda(B)|$ and $\sup_{B \subset A^c} |\lambda(B)|$ are finite, so is $\sup_B |\lambda(B)|$. Now let us complete the proof. Given a subset $A \in \mathcal{F}$ with $\sup_{B \subset A} |\lambda(B)| = \infty$, and any positive number N, there is a subset $A_1 \in \mathcal{F}$ with $A_1 \subset A$ such that $|\lambda(A_1)| \geq N$ and $\sup_{B \subset A_1} |\lambda(B)| = \infty$. This is obvious because if we pick a set $E \subset A$ with $|\lambda(E)|$ very large, then $\lambda(E^c)$ will be large as well. At least one of the two sets E and E^c will have the second property, and we can call it A_1. If we proceed by induction we have a sequence A_n that is $\downarrow$ and $|\lambda(A_n)| \to \infty$ that contradicts countable additivity. $\square$

LEMMA 4.2 *Given a subset $A \in \mathcal{F}$ with $\lambda(A) = \ell > 0$ there is a subset $\bar{A} \subset A$ that is totally positive with $\lambda(\bar{A}) \geq \ell$.*

PROOF: Let us define $m = \inf_{B \subset A} \lambda(B)$. Since the empty set is included, $m \leq 0$. If $m = 0$, then A is totally positive and we are done. So let us assume that $m < 0$. By the previous lemma $m > -\infty$.

Let us find $B_1 \subset A$ such that $\lambda(B_1) \leq \frac{m}{2}$. Then for $A_1 = A - B_1$ we have $A_1 \subset A$, $\lambda(A_1) \geq \ell$, and $\inf_{B \subset A_1} \lambda(B) \geq \frac{m}{2}$. By induction we can find A_k with $A \supset A_1 \supset \cdots \supset A_k \ldots$, $\lambda(A_k) \geq \ell$ for every k, and $\inf_{B \subset A_k} \lambda(A_k) \geq \frac{m}{2^k}$. Clearly, if we define $\bar{A} = \bigcap A_k$ which is the decreasing limit, $\bar{A}$ works. $\square$

THEOREM 4.3 (Hahn-Jordan Decomposition) *Given a countably additive signed measure λ on $(\Omega, \mathcal{F})$, it can be written always as $\lambda = \mu^+ - \mu^-$, the difference of two nonnegative measures. Moreover, μ^+ and μ^- may be chosen to be orthogonal, i.e., there are disjoint sets $\Omega_+, \Omega_- \in \mathcal{F}$ such that $\mu^+(\Omega_-) = \mu^-(\Omega_+) = 0$. In fact, Ω_+ and Ω_- can be taken to be subsets of Ω that are respectively totally positive and totally negative for λ; $\mu^\pm$ then become just the restrictions of λ to $\Omega_\pm$.*

PROOF: Totally positive sets are closed under countable unions, disjoint or not. Let us define $m^+ = \sup_A \lambda(A)$. If $m^+ = 0$, then $\lambda(A) \leq 0$ for all A and we can take $\Omega_+ = \emptyset$ and $\Omega_- = \Omega$, which works. Assume that $m^+ > 0$. There exist sets A_n with $\lambda(A) \geq m^+ - \frac{1}{n}$ and therefore totally positive subsets $\bar{A}_n$ of A_n with $\lambda(\bar{A}_n) \geq m^+ - \frac{1}{n}$. Clearly $\Omega_+ = \bigcup_n \bar{A}_n$ is totally positive and $\lambda(\Omega_+) = m^+$. It is easy to see that $\Omega_- = \Omega - \Omega_+$ is totally negative. $\mu^\pm$ can be taken to be the restriction of λ to $\Omega_\pm$. □

REMARK 4.2. If $\lambda = \mu^+ - \mu^-$ with μ^+ and μ^- orthogonal to each other, then they have to be the restrictions of λ to the totally positive and totally negative sets for λ, and such a representation for λ is unique. It is clear that in general the representation is not unique because we can add a common μ to both μ^+ and μ^- and the μ will cancel when we compute $\lambda = \mu^+ - \mu^-$.

REMARK 4.3. If μ is a nonnegative measure and we define λ by

$$\lambda(A) = \int_A f(\omega) d\mu = \int \chi_A(\omega) f(\omega) d\mu$$

where f is an integrable function, then λ is a countably additive signed measure and $\Omega_+ = \{\omega : f(\omega) > 0\}$ and $\Omega_- = \{\omega : f(\omega) < 0\}$. If we define $f^\pm(\omega)$ as the positive and negative parts of f, then

$$\mu^\pm(A) = \int_A f^\pm(\omega) d\mu \,.$$

The signed measure λ that was constructed in the preceding remark enjoys a very special relationship to μ. For any set A with $\mu(A) = 0$, $\lambda(A) = 0$ because the integrand $\chi_A(\omega) f(\omega)$ is 0 for μ-almost all ω and for all practical purposes is a function that vanishes identically.

DEFINITION 4.2 A signed measure λ is said to be *absolutely continuous with respect to a nonnegative measure* μ, $\lambda \ll \mu$ in symbols, if whenever $\mu(A)$ is zero for a set $A \in \mathcal{F}$ it is also true that $\lambda(A) = 0$.

THEOREM 4.4 (Radon-Nikodym Theorem) *If $\lambda \ll \mu$, then there is an integrable function $f(\omega)$ such that*

(4.1) $$\lambda(A) = \int_A f(\omega) d\mu$$

4.1. CONDITIONING

for all $A \in \mathcal{F}$. The function f is uniquely determined almost everywhere and is called the Radon-Nikodym derivative of λ with respect to μ. It is denoted by

$$f(\omega) = \frac{d\lambda}{d\mu}.$$

PROOF: The proof depends on the decomposition theorem. We saw that if relation (4.1) holds, then $\Omega_+ = \{\omega : f(\omega) > 0\}$. If we define $\lambda_a = \lambda - a\mu$, then λ_a is a signed measure for every real number a. Let us define $\Omega(a)$ to be the totally positive subset of λ_a. These sets are only defined up to sets of measure 0, and we can only handle a countable number of sets of measure 0 at one time. So it is prudent to restrict a to the set Q of rational numbers. Roughly speaking, $\Omega(a)$ will be the sets $f(\omega) > a$, and we will try to construct f from the sets $\Omega(a)$ by the definition

$$f(\omega) = [\sup a \in Q : \omega \in \Omega(a)].$$

The plan is to check that the function $f(\omega)$ defined above works. Since λ_a is getting more negative as a increases, $\Omega(a)$ is $\downarrow$ as $a \uparrow$. There is trouble with sets of measure 0 for every comparison between two rationals a_1 and a_2. Collect all such troublesome sets (only a countable number and throw them away). In other words, we may assume without loss of generality that $\Omega(a_1) \subset \Omega(a_2)$ whenever $a_1 > a_2$. Clearly,

$$\{\omega : f(\omega) > x\} = \{\omega : \omega \in \Omega(y) \text{ for some rational } y > x\} = \bigcup_{\substack{y > x \\ y \in Q}} \Omega(y),$$

and this makes f measurable. If $A \subset \bigcap_a \Omega(a)$, then $\lambda(A) - a\mu(A) \geq 0$ for all A. If $\mu(A) > 0$, $\lambda(A)$ has to be infinite, which is not possible. Therefore $\mu(A)$ has to be zero and by absolute continuity $\lambda(A) = 0$ as well. On the other hand, if $A \cap \Omega(a) = \emptyset$ for all a, then $\lambda(A) - a\mu(A) \leq 0$ for all a and again if $\mu(A) > 0$, $\lambda(A) = -\infty$, which is not possible either. Therefore $\mu(A)$ and, by absolute continuity, $\lambda(A)$ are zero. This proves that $f(\omega)$ is finite almost everywhere with respect to both λ and μ.

Let us take two real numbers $a < b$ and consider $E_{a,b} = \{\omega : a \leq f(\omega) \leq b\}$. It is clear that the set $E_{a,b}$ is in $\Omega(a')$ and $\Omega^c(b')$ for any $a' < a$ and $b' > b$. Therefore for any set $A \subset E_{a,b}$ by letting a' and b' tend to a and b

$$a\mu(A) \leq \lambda(A) \leq b\mu(A).$$

Now we are essentially done. Let us take a grid $\{n\,h\}$ and consider $E_n = \{\omega : nh \leq f(\omega) < (n+1)h\}$ for $-\infty < n < \infty$. Then, for any $A \in \mathcal{F}$ and each n,

$$\lambda(A \cap E_n) - h\mu(A \cap E_n) \leq nh\mu(A \cap E_n) \leq \int_{A \cap E_n} f(\omega)d\mu$$

$$\leq (n+1)h\mu(A \cap E_n) \leq \lambda(A \cap E_n) + h\mu(A \cap E_n).$$

Summing over n we have

$$\lambda(A) - h\mu(A) \leq \int_A f(\omega)d\mu \leq \lambda(A) + h\mu(A),$$

proving the integrability of f and, if we let $h \to 0$, establishing

$$\lambda(A) = \int_A f(\omega)d\mu \quad \text{for all } A \in \mathcal{F}.$$

□

REMARK 4.4 (Uniqueness). If we have two choices of f, say f_1 and f_2, their difference $g = f_1 - f_2$ satisfies

$$\int_A g(\omega)d\mu = 0 \quad \text{for all } A \in \mathcal{F}.$$

If we take $A_\varepsilon = \{\omega : g(\omega) \geq \varepsilon\}$, then $0 \geq \varepsilon\mu(A_\varepsilon)$ and this implies $\mu(A_\varepsilon) = 0$ for all $\varepsilon > 0$ or $g(\omega) \leq 0$ almost everywhere with respect to μ. A similar argument establishes $g(\omega) \geq 0$ almost everywhere with respect to μ. Therefore, $g = 0$ a.e. μ, proving uniqueness.

EXERCISE 4.1. If f and g are two integrable functions, measurable with respect to a σ-field $\mathcal{B}$, and if $\int_A f(\omega)dP = \int_A g(\omega)dP$ for all sets $A \in \mathcal{B}_0$, a field that generates the σ-field $\mathcal{B}$, then $f = g$ a.e. P.

EXERCISE 4.2. If $\lambda(A) \geq 0$ for all $A \in \mathcal{F}$, prove that $f(\omega) \geq 0$ almost everywhere.

EXERCISE 4.3. If Ω is a countable set and $\mu(\{\omega\}) > 0$ for each single point set, prove that any measure λ is absolutely continuous with respect to μ and calculate the Radon-Nikodym derivative.

EXERCISE 4.4. Let $F(x)$ be a distribution function on the line with $F(0) = 0$ and $F(1) = 1$ so that the probability measure α corresponding to it lives on the interval $[0, 1]$. If $F(x)$ satisfies a Lipschitz condition

$$|F(x) - F(y)| \leq A|x - y|,$$

then prove that $\alpha \ll m$ where m is the Lebesgue measure on $[0, 1]$. Show also that $0 \leq \frac{d\alpha}{dm} \leq A$ almost surely.

EXERCISE 4.5. If ν, λ and μ are three nonnegative measures such that $\nu \ll \lambda$ and $\lambda \ll \mu$, then show that $\nu \ll \mu$ and

$$\frac{d\nu}{d\mu} = \frac{d\nu}{d\lambda}\frac{d\lambda}{d\mu} \quad \text{a.e.}$$

EXERCISE 4.6. If λ and μ are nonnegative measures with $\lambda \ll \mu$ and $\frac{d\lambda}{d\mu} = f$, then show that g is integrable with respect to λ if and only if gf is integrable with respect to μ and

$$\int g(\omega)d\lambda = \int g(\omega)f(\omega)d\mu.$$

EXERCISE 4.7. Given two nonnegative measures λ and μ, λ is said to be *uniformly absolutely continuous* with respect to μ on $\mathcal{F}$ if for any $\varepsilon > 0$ there exists a $\delta > 0$ such that for any $A \in \mathcal{F}$ with $\mu(A) < \delta$ it is true that $\lambda(A) < \varepsilon$. Use the

Radon-Nikodym theorem to show that absolute continuity on a σ-field $\mathcal{F}$ implies uniform absolute continuity. If $\mathcal{F}_0$ is a field that generates the σ-field $\mathcal{F}$, show by an example that absolute continuity on $\mathcal{F}_0$ does not imply absolute continuity on $\mathcal{F}$. Show, however, that uniform absolute continuity on $\mathcal{F}_0$ implies uniform absolute continuity and therefore absolute continuity on $\mathcal{F}$.

EXERCISE 4.8. If F is a distribution function on the line, show that it is absolutely continuous with respect to the Lebesgue measure on the line if and only if for any $\varepsilon > 0$ there exists a $\delta > 0$ such that for an arbitrary finite collection of disjoint intervals $I_j = [a_j, b_j]$ with $\sum_j |b_j - a_j| < \delta$ it follows that $\sum_j [F(b_j) - F(a_j)] \leq \varepsilon$.

4.2. Conditional Expectation

In the Radon-Nikodym theorem, if $\lambda \ll \mu$ are two probability distributions on $(\Omega, \mathcal{F})$, we define the Radon-Nikodym derivative $f(\omega) = \frac{d\lambda}{d\mu}$ as an $\mathcal{F}$-measurable function such that $\lambda(A) = \int_A f(\omega) d\mu$ for all $A \in \mathcal{F}$. If $\Sigma \subset \mathcal{F}$ is a sub σ-field, the absolute continuity of λ with respect to μ on Σ is clearly implied by the absolute continuity of λ with respect to μ on $\mathcal{F}$. We can therefore apply the Radon-Nikodym theorem on the measurable space (Ω, Σ), and we will obtain a new Radon-Nikodym derivative

$$g(\omega) = \frac{d\lambda}{d\mu} = \frac{d\lambda}{d\mu}\bigg|_\Sigma$$

such that $\lambda(A) = \int_A g(\omega) d\mu$ for all $A \in \Sigma$ and g is Σ-measurable. Since the old function $f(\omega)$ was only $\mathcal{F}$-measurable, in general, it cannot be used as the Radon-Nikodym derivative for the sub σ-field Σ. Now, if f is an integrable function on $(\Omega, \mathcal{F}, \mu)$ and $\Sigma \subset \mathcal{F}$ is a sub σ-field, we can define λ on $\mathcal{F}$ by

$$\lambda(A) = \int_A f(\omega) d\mu \quad \text{for all } A \in \mathcal{F}$$

and recalculate the Radon-Nikodym derivative g for Σ and g will be a Σ-measurable, integrable function such that

$$\lambda(A) = \int_A g(\omega) d\mu \quad \text{for all } A \in \Sigma.$$

In other words, g is the perfect candidate for the conditional expectation

$$g(\omega) = E\{f(\cdot) \mid \Sigma\}.$$

We have therefore proved the existence of the conditional expectation.

THEOREM 4.5 *The conditional expectation as a mapping of $f \to g$ has the following properties*:

(i) *If* $g = E\{f \mid \Sigma\}$, *then* $E[g] = E[f]$. $E[1 \mid \Sigma] = 1$ *a.e.*
(ii) *If f is nonnegative, then* $g = E\{f \mid \Sigma\}$ *is almost surely nonnegative.*
(iii) *The map is linear. If a_1 and a_2 are constants*

$$E\{a_1 f_1 + a_2 f_2 \mid \Sigma\} = a_1 E\{f_1 \mid \Sigma\} + a_2 E\{f_2 \mid \Sigma\} \quad \text{a.e.}$$

(iv) If $g = E\{f \mid \Sigma\}$, then

$$\int |g(\omega)|d\mu \leq \int |f(\omega)|d\mu.$$

(v) If h is a bounded Σ-measurable function, then

$$E\{fh \mid \Sigma\} = hE\{f \mid \Sigma\} \quad \text{a.e.}$$

(vi) If $\Sigma_2 \subset \Sigma_1 \subset \mathcal{F}$, then

$$E[f \mid \Sigma_2] = E[E[f \mid \Sigma_1] \mid \Sigma_2].$$

(vii) Jensen's Inequality. If $\phi(x)$ is a convex function of x and $g = E\{f \mid \Sigma\}$, then

(4.2) $$E\{\phi(f(\omega)) \mid \Sigma\} \geq \phi(g(\omega)) \quad \text{a.e.}$$

and if we take expectations

$$E[\phi(f)] \geq E[\phi(g)].$$

PROOF: (i)–(iii) are obvious. For (iv) we note that if $d\lambda = f \, d\mu$,

$$\int |f|d\mu = \sup_{A \in \mathcal{F}} \lambda(A) - \inf_{A \in \mathcal{F}} \lambda(A),$$

and if we replace $\mathcal{F}$ by a sub σ-field Σ, the right-hand side is decreased. (v) is obvious if h is the indicator function of a set A in Σ. To go from indicator functions to simple functions to bounded measurable functions is routine. (vi) is an easy consequence of the definition and is left as an exercise. Finally, (vii) corresponds to Theorem 1.7 proved for ordinary expectations and is proved analogously. We note that if $f_1 \geq f_2$, then $E\{f_1 \mid \Sigma\} \geq E\{f_2 \mid \Sigma\}$ a.e. and, consequently, $E\{\max(f_1, f_2) \mid \Sigma\} \geq \max(g_1, g_2)$ a.e. where $g_i = E\{f_i \mid \Sigma\}$ for $i = 1, 2$. Since we can represent any convex function ϕ as $\phi(x) = \sup_a[ax - \psi(a)]$, limiting ourselves to rational a, we have only a countable set of functions to deal with, and

$$E\{\phi(f) \mid \Sigma\} = E\left\{\sup_a[af - \psi(a)] \mid \Sigma\right\}$$
$$\geq \sup_a[aE\{f \mid \Sigma\} - \psi(a)] = \sup_a[ag - \psi(a)] = \phi(g) \quad \text{a.e.}$$

and after taking expectations

$$E[\phi(f)] \geq E[\phi(g)].$$

□

REMARK 4.5. Conditional expectation is a form of averaging, i.e., it is linear, takes constants into constants, and preserves nonnegativity. Jensen's inequality is now a consequence of convexity.

In a somewhat more familiar context, if $\mu = \lambda_1 \times \lambda_2$ is a product measure on $(\Omega, \mathcal{F}) = (\Omega_1 \times \Omega_2, \mathcal{F}_1 \times \mathcal{F}_2)$ and we take $\Sigma = \{A \times \Omega_2 : A \in \mathcal{F}_1\}$, then for any function $f(\omega) = f(\omega_1, \omega_2)$, $E[f(\cdot) \mid \Sigma] = g(\omega)$ where $g(\omega) = g(\omega_1)$ is given by

$$g(\omega_1) = \int_{\Omega_2} f(\omega_1, \omega_2) d\lambda_2$$

so that the conditional expectation is just integrating the unwanted variable ω_2. We can go one step more. If $\phi(x, y)$ is the joint density on $\mathbb{R}^2$ of two random variables X and Y (with respect to the Lebesgue measure on $\mathbb{R}^2$), and $\psi(x)$ is the marginal density of X given by

$$\psi(x) = \int_{-\infty}^{\infty} \phi(x, y) dy,$$

then for any integrable function $f(x, y)$

$$E[f(X, Y) \mid X] = E[f(\cdot, \cdot) \mid \Sigma] = \frac{\int_{-\infty}^{\infty} f(x, y) \phi(x, y) dy}{\psi(x)}$$

where Σ is the σ-field of vertical strips $A \times (-\infty, \infty)$ with a measurable horizontal base A.

EXERCISE 4.9. If f is already Σ-measurable, then $E[f \mid \Sigma] = f$. This suggests that the map $f \to g = E[f \mid \Sigma]$ is some sort of a projection. In fact, if we consider the Hilbert space $\mathbf{H} = L_2[\Omega, \mathcal{F}, \mu]$ of all $\mathcal{F}$-measurable square-integrable functions with an inner product

$$\langle f, g \rangle_\mu = \int fg \, d\mu,$$

then

$$\mathbf{H}_0 = L_2[\Omega, \Sigma, \mu] \subset \mathbf{H} = L_2[\Omega, \mathcal{F}, \mu]$$

and $f \to E[f \mid \Sigma]$ is seen to be the same as the orthogonal projection from $\mathbf{H}$ onto $\mathbf{H}_0$. Prove it.

EXERCISE 4.10. If $\mathcal{F}_1 \subset \mathcal{F}_2 \subset \mathcal{F}$ are two sub σ-fields of $\mathcal{F}$ and X is any integrable function, we can define $X_i = E[X \mid \mathcal{F}_i]$ for $i = 1, 2$. Show that $X_1 = E[X_2 \mid \mathcal{F}_1]$ a.e.

Conditional expectation is then the best nonlinear predictor if the loss function is the expected (mean) square error.

4.3. Conditional Probability

We now turn our attention to conditional probability. If we take $f = \chi_B(\omega)$, then $E[f \mid \Sigma] = P(\omega, B)$ is called the *conditional probability* of B given Σ. It is characterized by the property that it is Σ-measurable as a function of ω and

$$(\star) \quad \mu(A \cap B) = \int_A P(\omega, B) d\mu \quad \text{for any } A \in \Sigma.$$

82   4. DEPENDENT RANDOM VARIABLES

THEOREM 4.6 $P(\cdot, \cdot)$ has the following properties:
   (i) $P(\omega, \Omega) = 1$, $P(\omega, \varnothing) = 0$ a.e.
   (ii) For any $B \in \mathcal{F}$, $0 \leq P(\omega, B) \leq 1$ a.e.
   (iii) For any countable collection $\{B_j\}$ of disjoint sets in $\mathcal{F}$,

$$P\left(\omega, \bigcup_j B_j\right) = \sum_j P(\omega, B_j) \quad \text{a.e.}$$

   (iv) If $B \in \Sigma$, $P(\omega, B) = \chi_B(\omega)$ a.e.

PROOF: All four properties are easy consequences of properties of conditional expectations. Property (iii) perhaps needs an explanation. If $E[|f_n - f|] \to 0$ by the properties of conditional expectation $E[|E\{f_n \mid \Sigma\} - E\{f \mid \Sigma\}|] \to 0$. Property (iii) is an easy consequence of this. □

The problem with the above theorem is that every property is valid only almost everywhere. There are exceptional sets of measure 0 for each case. While each null set or a countable number of them can be ignored, we have an uncountable number of null sets and we would like a single null set outside which all the properties hold. This means constructing a good version of the conditional probability. It may not always be possible. If possible, such a version is called a *regular conditional probability*. The existence of such a regular version depends on the space $(\Omega, \mathcal{F})$ and the sub σ-field Σ being nice. If Ω is a complete separable metric space and $\mathcal{F}$ are its Borel sets, and if Σ is any countably generated sub σ-field of $\mathcal{F}$, then it is nice enough. We will prove it in the special case when $\Omega = [0, 1]$ is the unit interval and $\mathcal{F}$ are the Borel subsets $\mathcal{B}$ of $[0, 1]$. Σ can be any countably generated sub σ-field of $\mathcal{F}$.

REMARK 4.6 In fact, the case is not so special. There is a theorem, see [6], which states that if $(\Omega, \mathcal{F})$ is any complete separable metric space that has an uncountable number of points, then there is a one-to-one measurable map with a measurable inverse between $(\Omega, \mathcal{F})$ and $([0, 1], \mathcal{B})$. There is no loss of generality in assuming that $(\Omega, \mathcal{F})$ is just $([0, 1], \mathcal{B})$.

THEOREM 4.7 Let P be a probability distribution on $([0, 1], \mathcal{B})$. Let $\Sigma \subset \mathcal{B}$ be a sub σ-field. There exists a family of probability distributions Q_x on $([0, 1], \mathcal{B})$ such that for every $A \in \mathcal{B}$, $Q_x(A)$ is Σ-measurable and for every $\mathcal{B}$-measurable f,

(4.3) $$\int f(y) Q_x(dy) = E^P[f(\cdot) \mid \Sigma] \quad \text{a.e. } P.$$

If in addition Σ is countably generated, i.e., there is a field Σ_0 consisting of a countable number of Borel subsets of $[0, 1]$ such that the σ-field generated by Σ_0 is Σ, then

(4.4) $$Q_x(A) = \mathbf{1}_A(x) \quad \text{for all } A \in \Sigma.$$

PROOF: The trick is not to be too ambitious in the first place but try to construct the conditional expectations

$$Q(\omega, B) = E\{\chi_B(\omega) \mid \Sigma\}$$

4.3. CONDITIONAL PROBABILITY

only for sets B given by $B = (-\infty, x]$ for rational x. We call our conditional expectation, which is in fact, a conditional probability, $F(\omega, x)$. By the properties of conditional expectations for any pair of rationals $x < y$, there is a null set $E_{x,y}$ such that for $\omega \notin E_{x,y}$

$$F(\omega, x) \le F(\omega, y).$$

Moreover, for any rational $x < 0$, there is a null set N_x outside which $F(\omega, x) = 0$ and similar null sets N_x for $x > 1$ outside which $F(\omega, x) = 1$. If we collect all these null sets, of which there are only countably many, and take their union, we get a null set $N \in \Sigma$ such that for $\omega \notin N$ we have a family $F(\omega, x)$ defined for rational x that satisfies

$$F(\omega, x) \le F(\omega, y) \quad \text{if } x < y \text{ are rational},$$
$$F(\omega, x) = 0 \quad \text{for rational } x < 0,$$
$$F(\omega, x) = 1 \quad \text{for rational } x > 1,$$
$$P(A \cap [0, x]) = \int_A F(\omega, x) dP \quad \text{for all } A \in \Sigma.$$

For $\omega \notin N$ and real y we can define

$$G(\omega, y) = \lim_{\substack{x \downarrow y \\ x \text{ rational}}} F(\omega, x).$$

For $\omega \notin N$, G is a right-continuous nondecreasing function (distribution function) with $G(\omega, y) = 0$ for $y < 0$ and $G(\omega, y) = 1$ for $y \ge 1$. There is then a probability measure $\widehat{Q}(\omega, B)$ on the Borel subsets of $[0, 1]$ such that $\widehat{Q}(\omega, [0, y]) = G(\omega, y)$ for all y. $\widehat{Q}$ is our candidate for regular conditional probability. Clearly $\widehat{Q}(\omega, I)$ is Σ-measurable for all intervals I and by standard arguments will continue to be Σ-measurable for all Borel sets $B \in \mathcal{F}$. If we check that

$$P(A \cap [0, x]) = \int_A G(\omega, x) dP \quad \text{for all } A \in \Sigma \text{ for all } 0 \le x \le 1,$$

then

$$P(A \cap I) = \int_A \widehat{Q}(\omega, I) dP \quad \text{for all } A \in \Sigma \text{ for all intervals } I,$$

and by standard arguments this will extend to finite disjoint unions of half-open intervals that constitute a field and, finally, to the σ-field $\mathcal{F}$ generated by that field.

To verify that for all real y,

$$P(A \cap [0, y]) = \int_A G(\omega, y) dP \quad \text{for all } A \in \Sigma,$$

we start from

$$P(A \cap [0, x]) = \int_A F(\omega, x) dP \quad \text{for all } A \in \Sigma$$

valid for rational x and let $x \downarrow y$ through the rationals. From the countable additivity of P the left-hand side converges to $P(A \cap [0, y])$ and by the bounded convergence theorem, the right-hand side converges to $\int_A G(\omega, y) \, dP$ and we are done.

Finally, from the uniqueness of the conditional expectation, if $A \in \Sigma$

$$\widehat{Q}(\omega, A) = \chi_A(\omega)$$

provided $\omega \notin N_A$, which is a null set that depends on A. We can take a countable set Σ_0 of generators A that forms a field and get a single null set N such that if $\omega \notin N$

$$\widehat{Q}(\omega, A) = \chi_A(\omega)$$

for all $A \in \Sigma_0$. Since both sides are countably additive measures in A and since they agree on Σ_0, they have to agree on Σ as well. □

EXERCISE 4.11 (Disintegration Theorem). Let μ be a probability measure on the plane $\mathbb{R}^2$ with a marginal distribution α for the first coordinate. In other words, α is such that, for any f that is a bounded measurable function of x,

$$\int_{R^2} f(x) d\mu = \int_R f(x) d\alpha \,.$$

Show that there exists a family of measures β_x depending measurably on x such that $\beta_x[\{x\} \times \mathbb{R}] = 1$, i.e., β_x is supported on the vertical line through $(x, y) : y \in \mathbb{R}$ and $\mu = \int_\mathbb{R} \beta_x \, d\alpha$. The converse is of course easier. Given α and β_x we can construct a unique μ such that μ disintegrates as expected.

4.4. Markov Chains

One of the ways of generating a sequence of dependent random variables is to think of a system evolving in time. We have time points that are discrete, say $T = 0, 1, \ldots, N, \ldots$. The state of the system is described by a point x in the state space $\mathcal{X}$ of the system. The state space $\mathcal{X}$ comes with a natural σ-field of subsets $\mathcal{F}$. At time 0 the system is in a random state and its distribution is specified by a probability distribution μ_0 on $(\mathcal{X}, \mathcal{F})$. At successive times $T = 1, 2, \ldots$, the system changes its state and given the past history $(x_0, x_1, \ldots, x_{k-1})$ of the states of the system at times $T = 0, 1, \ldots, k-1$ the probability that system finds itself at time k in a subset $A \in \mathcal{F}$ is given by $\pi_k(x_0, x_1, \ldots, x_{k-1}; A)$. For each $(x_0, x_1, \ldots, x_{k-1})$, π_k defines a probability measure on $(\mathcal{X}, \mathcal{F})$ and for each $A \in \mathcal{F}$, $\pi_k(x_0, x_1, \ldots, x_{k-1}; A)$ is assumed to be a measurable function of $(x_0, x_1, \ldots, x_{k-1})$ on the space $(\mathcal{X}^k, \mathcal{F}^k)$, which is the product of k copies of the space $(\mathcal{X}, \mathcal{F})$ with itself. We can inductively define measures μ_k on $(\mathcal{X}^{k+1}, \mathcal{F}^{k+1})$ that describe the probability distribution of the entire history $(x_0, x_1, \ldots, x_k)$ of the system through time k. To go from μ_{k-1} to μ_k, we think of $(\mathcal{X}^{k+1}, \mathcal{F}^{k+1})$ as the product of $(\mathcal{X}^k, \mathcal{F}^k)$ with $(\mathcal{X}, \mathcal{F})$ and construct on $(\mathcal{X}^{k+1}, \mathcal{F}^{k+1})$ a probability measure with marginal μ_{k-1} on $(\mathcal{X}^k, \mathcal{F}^k)$ and conditionals $\pi_k(x_0, x_1, \ldots, x_{k-1}; \cdot)$ on the fibers $(x_0, x_1, \ldots, x_{k-1}) \times \mathcal{X}$. This will define μ_k and the induction can proceed. We may stop at some finite terminal time N or go on indefinitely. If we do go on indefinitely, we will have a consistent family of finite-dimensional distributions $\{\mu_k\}$ on $(\mathcal{X}^{k+1}, \mathcal{F}^{k+1})$, and we may try to use Kolmogorov's theorem to construct a probability measure P on the space $(\mathcal{X}^\infty, \mathcal{F}^\infty)$ of sequences $\{x_j : j \geq 0\}$ representing the total evolution of the system for all times.

REMARK 4.7. Kolmogorov's theorem requires some assumptions on $(\mathcal{X}, \mathcal{F})$ that are satisfied if $\mathcal{X}$ is a complete separable metric space and $\mathcal{F}$ is the Borel sets. However, in the present context, there is a result known as Tulcea's theorem (see [8]) that proves the existence of a P on $(\mathcal{X}^\infty, \mathcal{F}^\infty)$ for any choice of $(\mathcal{X}, \mathcal{F})$, exploiting the fact that the consistent family of finite-dimensional distributions μ_k arises from well-defined successive regular conditional probability distributions.

An important subclass is generated when the transition probability depends on the past history only through the current state. In other words,
$$\pi_k(x_0, x_1, \ldots, x_{k-1}; \cdot) = \pi_{k-1,k}(x_{k-1}; \cdot).$$

In such a case the process is called a *Markov process* with transition probabilities $\pi_{k-1,k}(\cdot, \cdot)$. An even smaller subclass arises when we demand that $\pi_{k-1,k}(\cdot, \cdot)$ be the same for different values of k. In this case, a single transition probability $\pi(x, A)$ and the initial distribution μ_0 determine the entire process, i.e., the measure P on $(\mathcal{X}^\infty, \mathcal{F}^\infty)$. Such processes are called *time-homogeneous* Markov processes or Markov processes with stationary transition probabilities.

Chapman-Kolmogorov Equations. If we have the transition probabilities $\pi_{k,k+1}$ of transition from time k to $k+1$ of a Markov chain, it is possible to obtain directly the transition probabilities from time k to $k+\ell$ for any $\ell \geq 2$. We do it by induction on ℓ. Define

(4.5) $$\pi_{k,k+\ell+1}(x, A) = \int_\mathcal{X} \pi_{k,k+\ell}(x, dy) \pi_{k+\ell, k+\ell+1}(y, A)$$

or, equivalently, in a more direct fashion

$$\pi_{k,k+\ell+1}(x, A) = \int_\mathcal{X} \cdots \int_\mathcal{X} \pi_{k,k+1}(x, dy_{k+1}) \cdots \pi_{k+\ell, k+\ell+1}(y_{k+\ell}, A).$$

THEOREM 4.8 *The transition probabilities $\pi_{k,m}(\cdot, \cdot)$ satisfy the relations*

(4.6) $$\pi_{k,n}(x, A) = \int_\mathcal{X} \pi_{k,m}(x, dy) \pi_{m,n}(y, A) \quad \text{for any } k < m < n$$

and for the Markov process defined by the one-step transition probabilities $\pi_{k,k+1}(\cdot, \cdot)$,

$$P[x_n \in A \mid \Sigma_m] = \pi_{m,n}(x_m, A) \quad \text{a.e. for any } n > m$$

where Σ_m is the σ-field of past history up to time m generated by the coordinates $x_0, x_1, \ldots, x_m$.

PROOF: The identity (4.5) is obvious. The multiple integral can be carried out by iteration in any order, and after enough variables are integrated we get our identity. To prove that the conditional probabilities are given by the right formula, we need to establish

$$P[\{x_n \in A\} \cap B] = \int_B \pi_{m,n}(x_m, A) dP \quad \text{for all } B \in \Sigma_m \text{ and } A \in \mathcal{F}.$$

We write

$$P[\{x_n \in A\} \cap B]$$

$$= \int_{\{x_n \in A\} \cap B} dP$$

$$= \int \cdots \int_{\{x_n \in A\} \cap B} d\mu(x_0) \pi_{0,1}(x_0, dx_1) \cdots \pi_{m-1,m}(x_{m-1}, dx_m)$$

$$\pi_{m,m+1}(x_m, dx_{m-1}) \cdots \pi_{n-1,n}(x_{n-1}, dx_n)$$

$$= \int \cdots \int_B d\mu(x_0) \pi_{0,1}(x_0, dx_1) \cdots \pi_{m-1,m}(x_{m-1}, dx_m)$$

$$\pi_{m,m+1}(x_m, dx_{m-1}) \cdots \pi_{n-1,n}(x_{n-1}, A)$$

$$= \int \cdots \int_B d\mu(x_0) \pi_{0,1}(x_0, dx_1) \cdots \pi_{m-1,m}(x_{m-1}, dx_m) \pi_{m,n}(x_m, A)$$

$$= \int_B \pi_{m,n}(x_m, A) dP$$

and we are done. □

REMARK 4.8. *If the chain has stationary transition probabilities, then the transition probabilities $\pi_{m,n}(x, dy)$ from time m to time n depend only on the difference $k = n - m$ and are given by what are usually called the k-step transition probabilities.* They are defined inductively by

$$\pi^{(k+1)}(x, A) = \int_X \pi^{(k)}(x, dy) \pi(y, A)$$

and satisfy the Chapman-Kolmogorov equations

$$\pi^{(k+\ell)}(x, A) = \int_X \pi^{(k)}(x, dy) \pi^{(\ell)}(y, A) = \int_X \pi^{(\ell)}(x, dy) \pi^{(k)}(y, A).$$

Suppose we have a probability measure P on the product space $X \times Y \times Z$ with the product σ-field. The Markov property in this context refers to equality

(4.7) $$E^P[g(z) \mid \Sigma_{x,y}] = E^P[g(z) \mid \Sigma_y] \quad \text{a.e. } P$$

for bounded measurable functions g on Z, where we have used $\Sigma_{x,y}$ to denote the σ-field generated by projection onto $X \times Y$ and Σ_y the corresponding σ-field generated by projection onto Y. The Markov property in the reverse direction is the similar condition for bounded measurable functions f on X,

(4.8) $\qquad E^P[f(x) \mid \Sigma_{y,z}] = E^P[f(x) \mid \Sigma_y]$ a.e. P.

They look different. But they are both equivalent to the symmetric condition

(4.9) $\qquad E^P[f(x)g(z) \mid \Sigma_y] = E^P[f(x) \mid \Sigma_y] E^P[g(z) \mid \Sigma_y]$ a.e. P,

which says that given the present, the past and future are conditionally independent. In view of the symmetry, it is sufficient to prove the following:

THEOREM 4.9 *For any P on $(X \times Y \times Z)$ the relations (4.7) and (4.9) are equivalent.*

PROOF: Let us fix f and g. Let us denote the common value in (4.7) by $\hat{g}(y)$. Then

$$\begin{aligned}
E^P[f(x)g(z) \mid \Sigma_y] &= E^P\big[E^P[f(x)g(z) \mid \Sigma_{x,y}] \mid \Sigma_y\big] \quad \text{a.e. } P \\
&= E^P\big[f(x) E^P[g(z) \mid \Sigma_{x,y}] \mid \Sigma_y\big] \quad \text{a.e. } P \\
&= E^P[f(x)\hat{g}(y) \mid \Sigma_y] \quad \text{a.e. } P \quad \text{(by (4.7))} \\
&= E^P[f(x) \mid \Sigma_y]\hat{g}(y) \quad \text{a.e. } P \\
&= E^P[f(x) \mid \Sigma_y] E^P[g(z) \mid \Sigma_y] \quad \text{a.e. } P,
\end{aligned}$$

which is (4.9). Conversely, we assume (4.9) and denote by $\bar{g}(x,y)$ and $\hat{g}(y)$ the expressions on the left and right side of (4.7). Let $b(y)$ be a bounded measurable function on Y.

$$\begin{aligned}
E^P[f(x)b(y)\bar{g}(x,y)] &= E^P[f(x)b(y)g(z)] \\
&= E^P\big[b(y) E^P[f(x)g(z) \mid \Sigma_y]\big] \\
&= E^P\big[b(y)\{E^P[f(x) \mid \Sigma_y]\}\{E^P[g(z) \mid \Sigma_y]\}\big] \\
&= E^P\big[b(y)\{E^P[f(x) \mid \Sigma_y]\}\hat{g}(y)\big] \\
&= E^P[f(x)b(y)\hat{g}(y)].
\end{aligned}$$

Since f and b are arbitrary, this implies that $\bar{g}(x,y) = \hat{g}(y)$ a.e. P. □

Let us look at some examples.

(1) Suppose we have an urn containing a certain number of balls (nonzero), some red and others green. A ball is drawn at random and its color is noted. Then it is returned to the urn along with an extra ball of the same color. Then a new ball is drawn at random and the process continues ad infinitum. The current state of the system can be characterized by two integers r and g such that $r+g \geq 1$. The initial state of the system is some r_0, g_0 with $r_0 + g_0 \geq 1$. The system can go from (r,g) to either $(r+1,g)$ with probability $\frac{r}{r+g}$ or to $(r,g+1)$ with probability $\frac{g}{r+g}$. This is clearly an example of a Markov chain with stationary transition probabilities.

(2) Consider a queue for service in a store. Suppose at each of the times $1, 2, \ldots$, a random number of new customers arrive and join the queue. If the queue is nonempty at some time, then exactly one customer will be served and will leave the queue at the next time point. The distribution of the number of new arrivals is specified by $\{p_j : j \geq 0\}$ where p_j is the probability that exactly j new customers arrive at a given time. The number of new arrivals at distinct times are assumed to be independent. The queue length is a Markov chain on the state space $\mathcal{X} = \{0, 1, \ldots\}$ of nonnegative integers. The transition probabilities $\pi(i, j)$ are given by $\pi(0, j) = p_j$ because there is no service and nobody in the queue to begin with and all the new arrivals join the queue. On the other hand, $\pi(i, j) = p_{j-i+1}$ if $j + 1 \geq i \geq 1$ because one person leaves the queue after being served.

(3) Consider a reservoir into which water flows. The amount of additional water flowing into the reservoir on any given day is random and has a distribution α on $[0, \infty)$. The demand is also random for any given day, with a probability distribution β on $[0, \infty)$. We may also assume that the inflows and demands on successive days are random variables ξ_n and η_n that have α and β for their common distributions and are all mutually independent. We may wish to assume a percentage loss due to evaporation. In any case the storage levels on successive days have a recurrence relation
$$S_{n+1} = [(1 - p)S_n + \xi_n - \eta_n]^+;$$
p is the loss and we have put the condition that the outflow is the demand unless the stored amount is less than the demand, in which case the outflow is the available quantity. The current amount in storage is a Markov process with stationary transition probabilities.

(4) Let $X_1, X_2, \ldots, X_n, \ldots$ be a sequence of independent random variables with a common distribution α. Let $S_n = Y + X_1 + + X_2 + \cdots + X_n$ for $n \geq 1$ with $S_0 = Y$ where Y is a random variable independent of $X_1, X_2, \ldots, X_n, \ldots$ with distribution μ. Then S_n is a Markov chain on $\mathbb{R}$ with one-step transition probability $\pi(x, A) = \alpha(A - x)$ and initial distribution μ. The n-step transition probability is $\alpha^n(A - x)$ where α^n is the n-fold convolution of α. This is often referred to as a *random walk*.

The last two examples can be described by models of the type
$$x_n = f(x_{n-1}, \xi_n)$$
where x_n is the current state and ξ_n is some random external disturbance. The ξ_n are assumed to be independent and identically distributed. They could have two components like inflow and demand. The new state is a deterministic function of the old state and the noise.

EXERCISE 4.12. Verify that the first two examples can be cast in the above form. In fact, there is no loss of generality in assuming that the ξ_j are mutually independent random variables having as common distribution the uniform distribution on the interval [0, 1].

Given a Markov chain with stationary transition probabilities $\pi(x, dy)$ on a state space $(\mathcal{X}, \mathcal{F})$, the behavior of $\pi^{(n)}(x, dy)$ for large n is an important and natural question. In the best situation of independent random variables $\pi^{(n)}(x, A) = \mu(A)$ are independent of x as well as n. Hopefully, after a long time the chain will "forget" its origins and $\pi^{(n)}(x, \cdot) \to \mu(\cdot)$, in some suitable sense, for some μ that does not depend on x. If that happens, then from the relation

$$\pi^{(n+1)}(x, A) = \int \pi^{(n)}(x, dy)\pi(y, A)$$

we conclude

$$\mu(A) = \int \pi(y, A) d\mu(y) \quad \text{for all } A \in \mathcal{F}.$$

Measures that satisfy the above property, abbreviated as $\mu\pi = \mu$, are called *invariant measures* for the Markov chain. If we start with the initial distribution μ which is invariant, then the probability measure P has μ as marginal at every time. In fact, P is stationary, i.e., invariant with respect to time translation, and can be extended to a stationary process where time runs from $-\infty$ to $+\infty$.

4.5. Stopping Times and Renewal Times

One of the important notions in the analysis of Markov chains is the idea of stopping times and renewal times. A function

$$\tau(\omega) : \Omega \to \{n : n \geq 0\}$$

is a random variable defined on the set $\Omega = \mathcal{X}^\infty$ such that for every $n \geq 0$ the set $\{\omega : \tau(\omega) = n\}$ (or equivalently for each $n \geq 0$ the set $\{\omega : \tau(\omega) \leq n\}$) is measurable with respect to the σ-field $\mathcal{F}_n$ generated by $X_j : 0 \leq j \leq n$. It is not necessary that $\tau(\omega) < \infty$ for every ω. Such random variables τ are called *stopping times*.

Examples of stopping times are: *constant times $n \geq 0$, the first visit to a state x*, or *the second visit to a state x*. The important thing is that in order to decide if $\tau \leq n$, i.e., to know if whatever is supposed to happen did happen before time n, the chain need be observed only up to time n. Examples of τ that are not stopping times are easy to find. The last time a site is visited is not a stopping time, nor is the first time such that at the next time one is in a state x.

An important fact is that the Markov property extends to stopping times. Just as we have σ-fields $\mathcal{F}_n$ associated with constant times, we do have a σ-field $\mathcal{F}_\tau$ associated to any stopping time. This is the information we have when we observe the chain up to time τ. Formally,

$$\mathcal{F}_\tau = \{A : A \in \mathcal{F}^\infty \text{ and } A \cap \{\tau \leq n\} \in \mathcal{F}_n \text{ for each } n\}.$$

One can check from the definition that τ is $\mathcal{F}_\tau$-measurable and so is X_τ on the set $\tau < \infty$. If τ is the time of first visit to y, then τ is a stopping time and the event that the chain visits a state z before visiting y is $\mathcal{F}_\tau$-measurable.

LEMMA 4.10 (Strong Markov Property) *At any stopping time τ the Markov property holds in the sense that the conditional distribution of $X_{\tau+1}, \ldots, X_{\tau+n}, \ldots$ conditioned on $\mathcal{F}_\tau$ is the same as the original chain starting from the state $x = X_\tau$ on the set $\tau < \infty$. In other words,*

$$P_x\{X_{\tau+1} \in A_1, \ldots, X_{\tau+n} \in A_n | \mathcal{F}_\tau\} = \int_{A_1} \cdots \int_{A_n} \pi(X_\tau, dx_1) \cdots \pi(x_{n-1}, dx_n) \quad \text{a.e. on } \{\tau < \infty\}.$$

PROOF: Let $A \in \mathcal{F}_\tau$ be given with $A \subset \{\tau < \infty\}$. Then

$$P_x\{A \cap \{X_{\tau+1} \in A_1, \ldots, X_{\tau+n} \in A_n\}\}$$

$$= \sum_k P_x\{A \cap \{\tau = k\} \cap \{X_{k+1} \in A_1, \ldots, X_{k+n} \in A_n\}\}$$

$$= \sum_k \int_{A \cap \{\tau=k\}} \int_{A_1} \cdots \int_{A_n} \pi(X_k, dx_{k+1}) \cdots \pi(x_{k+n-1}, dx_{k+n}) dP_x$$

$$= \int_A \int_{A_1} \cdots \int_{A_n} \pi(X_\tau, dx_1) \cdots \pi(x_{n-1}, dx_n) dP_x.$$

We have used the fact that if $A \in \mathcal{F}_\tau$, then $A \cap \{\tau = k\} \in \mathcal{F}_k$ for every $k \geq 0$. □

REMARK 4.9. If $X_\tau = y$ a.e. with respect to P_x on the set $\tau < \infty$, then at time τ, when it is finite, the process starts afresh with no memory of the past and will have conditionally the same probabilities in the future as P_y. At such times the process renews itself and these times are called *renewal times*.

4.6. Countable State Space

From the point of view of analysis, a particularly simple situation is when the state space $\mathcal{X}$ is a countable set. It can be taken as the integers $\{x : x \geq 1\}$. Many applications fall in this category, and an understanding of what happens in this situation will tell us what to expect in general.

The one-step transition probability is a matrix $\pi(x, y)$ with nonnegative entries such that $\sum_y \pi(x, y) = 1$ for each x. Such matrices are called *stochastic matrices*. The n-step transition matrix is just the nth power of the matrix defined inductively by

$$\pi^{(n+1)}(x, y) = \sum_z \pi^{(n)}(x, z) \pi(z, y).$$

To be consistent one defines $\pi^{(0)}(x, y) = \delta_{x,y}$, which is 1 if $x = y$ and 0 otherwise. The problem is to analyze the behavior for large n of $\pi^{(n)}(x, y)$. A state x is said to *communicate* with a state y if $\pi^{(n)}(x, y) > 0$ for some $n \geq 1$. We will assume

4.6. COUNTABLE STATE SPACE

for simplicity that every state communicates with every other state. Such Markov chains are called *irreducible*.

Let us first limit ourselves to the study of irreducible chains. Given an irreducible Markov chain with transition probabilities $\pi(x, y)$, we define $f_n(x)$ as the probability of returning to x for the first time at the nth step, assuming that the chain starts from the state x. Using the convention that P_x refers to the measure on sequences for the chain starting from x and $\{X_j\}$ are the successive positions of the chain,

$$f_n(x) = P_x\{X_j \neq x \text{ for } 1 \leq j \leq n-1 \text{ and } X_n = x\}$$
$$= \sum_{\substack{y_1 \neq x \\ \vdots \\ y_{n-1} \neq x}} \pi(x, y_1)\pi(y_1, y_2)\cdots\pi(y_{n-1}, x).$$

Since $f_n(x)$ are probabilities of disjoint events, $\sum_n f_n(x) \leq 1$. The state x is called *transient* if $\sum_n f_n(x) < 1$ and *recurrent* if $\sum_n f_n(x) = 1$. The recurrent case is divided into two situations. If we denote by $\tau_x = \inf\{n \geq 1 : X_n = x\}$, the time of first visit to x, then recurrence is $P_x\{\tau_x < \infty\} = 1$. A recurrent state x is called

positive recurrent if $E^{P_x}\{\tau_x\} = \sum_{n \geq 1} n f_n(x) < \infty$ and

null recurrent if $E^{P_x}\{\tau_x\} = \sum_{n \geq 1} n f_n(x) = \infty$.

LEMMA 4.11 *If for a (not necessarily irreducible) chain starting from x, the probability of ever visiting y is positive, then so is the probability of visiting y before returning to x.*

PROOF: Assume that for the chain starting from x the probability of visiting y before returning to x is zero. But when it returns to x it starts afresh and so will not visit y until it returns again. This reasoning can be repeated and so the chain will have to visit x infinitely often before visiting y. But this will use up all the time and so it cannot visit y at all. □

LEMMA 4.12 *For an irreducible chain all states x are of the same type.*

PROOF: Let x be recurrent and y be given. Since the chain is irreducible, for some k, $\pi^{(k)}(x, y) > 0$. By the previous lemma, for the chain starting from x, there is a positive probability of visiting y before returning to x. After each successive return to x, the chain starts afresh and there is a fixed positive probability of visiting y before the next return to x. Since there are infinitely many returns to x, y will be visited infinitely many times as well. Or y is also a recurrent state.

We now prove that if x is positive recurrent, then so is y. We saw already that the probability $p = P_x\{\tau_y < \tau_x\}$ of visiting y before returning to x is positive. Clearly,

$$E^{P_x}\{\tau_x\} \geq P_x\{\tau_y < \tau_x\} E^{P_y}\{\tau_x\} \quad \text{and therefore} \quad E^{P_y}\{\tau_x\} \leq \frac{1}{p} E^{P_x}\{\tau_x\} < \infty.$$

On the other hand, we can write

$$E^{P_x}\{\tau_y\} \leq \int_{\tau_y < \tau_x} \tau_x \, dP_x + \int_{\tau_x < \tau_y} \tau_y \, dP_x$$

$$= \int_{\tau_y < \tau_x} \tau_x \, dP_x + \int_{\tau_x < \tau_y} \{\tau_x + E^{P_x}\{\tau_y\}\} dP_x$$

$$= \int_{\tau_y < \tau_x} \tau_x \, dP_x + \int_{\tau_x < \tau_y} \tau_x \, dP_x + (1-p) E^{P_x}\{\tau_y\}$$

$$= \int \tau_x \, dP_x + (1-p) E^{P_x}\{\tau_y\}$$

by the renewal property at the stopping time τ_x. Therefore

$$E^{P_x}\{\tau_y\} \leq \frac{1}{p} E^{P_x}\{\tau_x\}.$$

We also have

$$E^{P_y}\{\tau_y\} \leq E^{P_y}\{\tau_x\} + E^{P_x}\{\tau_y\} \leq \frac{2}{p} E^{P_x}\{\tau_x\},$$

proving that y is positive recurrent. □

Transient Case. We have the following theorem regarding transience:

THEOREM 4.13 *An irreducible chain is transient if and only if*

$$G(x, y) = \sum_{n=0}^{\infty} \pi^{(n)}(x, y) < \infty \quad \text{for all } x, y.$$

Moreover, for any two states x and y,

$$G(x, y) = f(x, y) G(y, y) \quad \text{and} \quad G(x, x) = \frac{1}{1 - f(x, x)}$$

where $f(x, y) = P_x\{\tau_y < \infty\}$.

PROOF: Each time the chain returns to x there is a probability $1 - f(x, x)$ of never returning. The number of returns then has the geometric distribution

$$P_x\{\text{exactly } n \text{ returns to } x\} = (1 - f(x, x)) f(x, x)^n,$$

and the expected number of returns is given by

$$\sum_{k=1}^{\infty} \pi^{(k)}(x, x) = \frac{f(x, x)}{1 - f(x, x)}.$$

The left-hand side comes from the calculation

$$E^{P_x} \sum_{k=1}^{\infty} \chi_{\{x\}}(X_k) = \sum_{k=1}^{\infty} \pi^{(k)}(x, x),$$

and the right-hand side from the calculation of the mean of a geometric distribution. Since we count the visit at time 0 as a visit to x, we add 1 to both sides to get our formula. If we want to calculate the expected number of visits to y when we start from x, first we have to get to y and the probability of that is $f(x, y)$. Then by the renewal property it is exactly the same as the expected number of visits to y starting from y, including the visit at time 0 and that equals $G(y, y)$. □

Before we study the recurrent behavior we need the notion of periodicity. For each state x let us define $D_x = \{n : \pi^{(n)}(x, x) > 0\}$ to be the set of times at which a return to x is possible if one starts from x. We define d_x to be the greatest common divisor of D_x.

LEMMA 4.14 *For any irreducible chain $d_x = d$ for all $x \in \mathcal{X}$ and for each x, D_x contains all sufficiently large multiples of d.*

PROOF: Let us define
$$D_{x,y} = \{n : \pi^{(n)}(x, y) > 0\}$$
so that $D_x = D_{x,x}$. By the Chapman-Kolmogorov equations
$$\pi^{(m+n)}(x, y) \geq \pi^{(m)}(x, z)\pi^{(n)}(z, y)$$
for every z, so that if $m \in D_{x,z}$ and $n \in D_{z,y}$, then $m + n \in D_{x,y}$. In particular, if $m, n \in D_x$ it follows that $m + n \in D_x$. Since any pair of states communicate with each other, given $x, y \in \mathcal{X}$, there are positive integers n_1 and n_2 such that $n_1 \in D_{x,y}$ and $n_2 \in D_{y,x}$. This implies that with the choice of $\ell = n_1 + n_2$, $n + \ell \in D_x$ whenever $n \in D_y$; similarly, $n + \ell \in D_y$ whenever $n \in D_x$. Since ℓ itself belongs to both D_x and D_y, both d_x and d_y divide ℓ. Suppose $n \in D_x$. Then $n + \ell \in D_y$ and therefore d_y divides $n + \ell$. Since d_y divides ℓ, d_y must divide n. Since this is true for every $n \in D_x$ and d_x is the greatest common divisor of D_x, d_y must divide d_x. Similarly, d_x must divide d_y. Hence $d_x = d_y$.

We now complete the proof of the lemma. Let d be the greatest common divisor of D_x. Then it is the greatest common divisor of a finite subset $n_1, n_2, \ldots, n_q$ of D_x, and there will exist integers $a_1, a_2, \ldots, a_q$ such that
$$a_1 n_1 + a_2 n_2 + \cdots + a_q n_q = d.$$
Some of the a's will be positive and others negative. Separating them out and remembering that all the n_i are divisible by d, we find two integers md and $(m+1)d$ such that they both belong to D_x. If now $n = kd$ with $k > m^2$, we can write $k = \ell m + r$ with a large $\ell \geq m$ and the remainder r is less than m.
$$kd = (\ell m + r)d = \ell md + r(m + 1)d - rmd$$
$$= (\ell - r)md + r(m + 1)d \in D_x$$
since $\ell \geq m > r$. □

REMARK 4.10. For an irreducible chain the common value d is called the *period* of the chain, and an irreducible chain with period $d = 1$ is called *aperiodic*.

The simplest example of a periodic chain is one with two states, and the chain shuttles back and forth between the two. $\pi(x, y) = 1$ if $x \neq y$ and 0 if $x = y$. A simple calculation yields $\pi^{(n)}(x, x) = 1$ if n is even and 0 otherwise. There is oscillatory behavior in n that persists. The main theorem for irreducible, aperiodic, recurrent chains is the following:

THEOREM 4.15 *Let $\pi(x, y)$ be the one-step transition probability for a recurrent aperiodic Markov chain, and let $\pi^{(n)}(x, y)$ be the n-step transition probabilities. If the chain is null recurrent, then*

$$\lim_{n \to \infty} \pi^{(n)}(x, y) = 0 \quad \text{for all } x, y.$$

If the chain is positive recurrent, then of course $E^{P_x}\{\tau_x\} = m(x) < \infty$ for all x, and in that case

$$\lim_{n \to \infty} \pi^{(n)}(x, y) = q(y) = \frac{1}{m(y)}$$

exists for all x and y and is independent of the starting point x, and $\sum_y q(y) = 1$.

The proof is based on

THEOREM 4.16 (Renewal Theorem) *Let $\{f_n : n \geq 1\}$ be a sequence of nonnegative numbers such that*

$$\sum_n f_n = 1, \quad \sum_n n f_n = m \leq \infty,$$

and the greatest common divisor of $\{n : f_n > 0\}$ is 1. Suppose that $\{p_n : n \geq 0\}$ are defined by $p_0 = 1$ and recursively

(4.10) $$p_n = \sum_{j=1}^{n} f_j p_{n-j}.$$

Then

$$\lim_{n \to \infty} p_n = \frac{1}{m},$$

where if $m = \infty$ the right-hand side is taken as 0.

PROOF: The proof is based on several steps.

Step 1. We have inductively $p_n \leq 1$. Let $a = \limsup_{n \to \infty} p_n$. We can choose a subsequence n_k such that $p_{n_k} \to a$. We can assume without loss of generality that $p_{n_k + j} \to q_j$ as $k \to \infty$ for all positive and negative integers j as well. Of course, the limit q_0 for $j = 0$ is a. In relation (4.10) we can pass to the limit along the subsequence and use the dominated convergence theorem to obtain

(4.11) $$q_n = \sum_{j=1}^{\infty} f_j q_{n-j}$$

valid for $-\infty < n < \infty$; in particular,

(4.12) $$q_0 = \sum_{j=1}^{\infty} f_j q_{-j}.$$

Step 2. Because $a = \limsup p_n$, we can conclude that $q_j \leq a$ for all j. If we denote by $S = \{n : f_n > 0\}$, then $q_{-k} = a$ for $k \in S$. We can then deduce from equation (4.11) that $q_{-k} = a$ for $k = k_1 + k_2$ with $k_1, k_2 \in S$. By repeating the same reasoning, $q_{-k} = a$ for $k = k_1 + k_2 + \cdots + k_\ell$. By Lemma 3.6, because the greatest common factor of the integers in S is 1, there is a k_0 such that for $k \geq k_0$ we have $q_{-k} = a$. We now apply the relation (4.11) again to conclude that $q_j = a$ for all positive as well as negative j.

Step 3. If we add up equation (4.10) for $n = 1, 2, \ldots, N$ we get

$$p_1 + p_2 + \cdots + p_N = (f_1 + f_2 + \cdots + f_N) + (f_1 + f_2 + \cdots + f_{N-1})p_1$$
$$+ \cdots + (f_1 + f_2 + \cdots + f_{N-k})p_k + \cdots + f_1 p_{N-1}.$$

If we let $T_j = \sum_{i=j}^{\infty} f_i$, we have $T_1 = 1$ and $\sum_{j=0}^{\infty} T_j = m$. We can now rewrite

$$\sum_{j=1}^{N} T_j p_{N-j+1} = \sum_{j=1}^{N} f_j.$$

Step 4. Because $p_{N-j} \to a$ for every j along the subsequence $N = n_k$, if $\sum_j T_j = m < \infty$, we can deduce from the dominated convergence theorem that $ma = 1$ and we conclude that

$$\limsup_{n \to \infty} p_n = \frac{1}{m}.$$

If $\sum_j T_j = \infty$, by Fatou's lemma $a = 0$. Exactly the same argument applies to lim inf and we conclude that

$$\liminf_{n \to \infty} p_n = \frac{1}{m}.$$

This concludes the proof of the renewal theorem. □

We now turn to

PROOF OF THEOREM 4.15: If we take a fixed $x \in \mathcal{X}$ and consider $f_n = P_x\{\tau_x = n\}$, then f_n and $p_n = \pi^{(n)}(x, x)$ are related by (4.10) and $m = E^{P_x}\{\tau_x\}$. In order to apply the renewal theorem, we need to establish that the greatest common divisor of $S = \{n : f_n > 0\}$ is 1. In general, if $f_n > 0$, so is p_n. So the greatest common divisor of S is always larger than that of $\{n : p_n > 0\}$. That does not help us because the greatest common divisor of $\{n : p_n > 0\}$ is 1. On the other hand, if $f_n = 0$ unless $n = kd$ for some k, the relation (4.10) can be used inductively to conclude that the same is true of p_n. Hence, both sets have the same greatest common divisor. We can now conclude that

$$\lim_{n \to \infty} \pi^{(n)}(x, x) = q(x) = \frac{1}{m(x)}.$$

On the other hand, if $f_n(x, y) = P_x\{\tau_y = n\}$, then

$$\pi^{(n)}(x, y) = \sum_{k=1}^{n} f_k(x, y) \pi^{(n-k)}(y, y).$$

and recurrence implies $\sum_{k+1}^{\infty} f_k(x, y) = 1$ for all x and y. Therefore,

$$\lim_{n \to \infty} \pi^{(n)}(x, y) = q(y) = \frac{1}{m(y)}$$

and is independent of x, the starting point. In order to complete the proof we have to establish that

$$Q = \sum_y q(y) = 1.$$

It is clear by Fatou's lemma that

$$\sum_y q(y) = Q \leq 1.$$

By letting $n \to \infty$ in the Chapman-Kolmogorov equation

$$\pi^{(n+1)}(x, y) = \sum_z \pi^n(x, z) \pi(z, y)$$

and using Fatou's lemma, we get

$$q(y) \geq \sum_z \pi(z, y) q(z).$$

Summing with respect to y we obtain

$$Q \geq \sum_{z,y} \pi(z, y) q(z) = Q$$

and equality holds in this relation. Therefore,

$$q(y) = \sum_z \pi(z, y) q(z)$$

for every y or $q(\cdot)$ is an invariant measure. By iteration

$$q(y) = \sum_z \pi^n(z, y) q(z),$$

and if we let $n \to \infty$ again, an application of the bounded convergence theorem yields

$$q(y) = Q\, q(y),$$

implying $Q = 1$ and we are done. □

Let us now consider an irreducible Markov chain with one-step transition probability $\pi(x, y)$ that is periodic with period $d > 1$. Let us choose and fix a reference point $x_0 \in \mathcal{X}$. For each $x \in \mathcal{X}$, let $D_{x_0,x} = \{n : \pi^{(n)}(x_0, x) > 0\}$.

LEMMA 4.17 *If $n_1, n_2 \in D_{x_0,x}$, then d divides $n_1 - n_2$.*

PROOF: Since the chain is irreducible, there is an m such that $\pi^{(m)}(x, x_0) > 0$. By the Chapman-Kolmogorov equations $\pi^{(m+n_i)}(x_0, x_0) > 0$ for $i = 1, 2$. Therefore, $m + n_i \in D_{x_0} = D_{x_0,x_0}$ for $i = 1, 2$. This implies that d divides both $m + n_1$ as well as $m + n_2$. Thus, d divides $n_1 - n_2$. □

4.6. COUNTABLE STATE SPACE

The residue modulo d of all the integers in $D_{x_0,x}$ are the same and equal some number $r(x)$ satisfying $0 \leq r(x) \leq d-1$. By definition $r(x_0) = 0$. Let us define $\mathcal{X}_j = \{x : r(x) = j\}$. Then $\{\mathcal{X}_j : 0 \leq j \leq d-1\}$ is a partition of $\mathcal{X}$ into disjoint sets with $x_0 \in \mathcal{X}_0$.

LEMMA 4.18 *If $x \in \mathcal{X}$, then $\pi^{(n)}(x,y) = 0$ unless $r(x) + n = r(y) \mod d$.*

PROOF: Suppose that $x \in \mathcal{X}$ and $\pi(x,y) > 0$. Then if $m \in D_{x_0,x}$, then $(m+1) \in D_{x_0,y}$. Therefore $r(x) + 1 = r(y) \mod d$. The proof can be completed by induction. The chain marches through $\{\mathcal{X}_j\}$ in a cyclical way from a state in $\mathcal{X}_j$ to one in $\mathcal{X}_{j+1}$. □

THEOREM 4.19 *Let $\mathcal{X}$ be irreducible and positive recurrent with period d. Then*

$$\lim_{\substack{n \to \infty \\ n+r(x)=r(y) \mod d}} \pi^{(n)}(x,y) = \frac{d}{m(y)}.$$

Of course,

$$\pi^{(n)}(x,y) = 0$$

unless $n + r(x) = r(y) \mod d$.

PROOF: If we replace π by $\tilde{\pi}$ where $\tilde{\pi}(x,y) = \pi^{(d)}(x,y)$, then $\tilde{\pi}(x,y) = 0$ unless both x and y are in the same $\mathcal{X}_j$. The restriction of $\tilde{\pi}$ to each $\mathcal{X}_j$ defines an irreducible aperiodic Markov chain. Since each time step under $\tilde{\pi}$ is actually d units of time, we can apply the earlier results and we will get for $x, y \in \mathcal{X}_j$ for some j,

$$\lim_{k \to \infty} \pi^{(kd)}(x,y) = \frac{d}{m(y)}.$$

We note that

$$\pi^{(n)}(x,y) = \sum_{1 \leq m \leq n} f_m(x,y) \pi^{(n-m)}(y,y),$$

$$f_m(x,y) = P_x\{\tau_y = m\} = 0 \quad \text{unless } r(x) + m = r(y) \mod d,$$

$$\pi^{(n-m)}(y,y) = 0 \quad \text{unless } n - m = 0 \mod d,$$

$$\sum_m f_m(x,y) = 1.$$

The theorem now follows. □

Suppose now we have a chain that is not irreducible. Let us collect all the transient states and call the set $\mathcal{X}_{\text{tr}}$. The complement consists of all the recurrent states and will be denoted by $\mathcal{X}_{\text{re}}$.

LEMMA 4.20 *If $x \in \mathcal{X}_{\text{re}}$ and $y \in \mathcal{X}_{\text{tr}}$, then $\pi(x,y) = 0$.*

PROOF: If x is a recurrent state and $\pi(x,y) > 0$, the chain will return to x infinitely often and each time there is a positive probability of visiting y. By the renewal property these are independent events and so y will be recurrent too. □

The set of recurrent states $\mathcal{X}_{\text{re}}$ can be divided into one or more equivalence classes according to the following procedure: Two recurrent states x and y are in the same equivalence class if $f(x, y) = P_x\{\tau_y < \infty\}$; the probability of ever visiting y starting from x is positive. Because of recurrence, if $f(x, y) > 0$, then $f(x, y) = f(y, x) = 1$. The restriction of the chain to a single equivalence class is irreducible and possibly periodic. Different equivalence classes could have different periods, some could be positive recurrent and others null recurrent.

We can combine all our observations into the following theorem:

THEOREM 4.21 *If y is transient, then $\sum_n \pi^{(n)}(x, y) < \infty$ for all x. If y is null recurrent (belongs to an equivalence class that is null recurrent), then $\pi^{(n)}(x, y) \to 0$ for all x, but $\sum_n \pi^{(n)}(x, y) = \infty$ if x is in the same equivalence class or $x \in \mathcal{X}_{\text{tr}}$ with $f(x, y) > 0$. In all other cases $\pi^{(n)}(x, y) = 0$ for all $n \geq 1$. If y is positive recurrent and belongs to an equivalence class with period d with $m(y) = E^{P_y}\{\tau_y\}$, then for a nontransient x, $\pi^{(n)}(x, y) = 0$ unless x is in the same equivalence class and $r(x) + n = r(y) \mod d$. In such a case,*

$$\lim_{\substack{n \to \infty \\ r(x)+n=r(y) \bmod d}} \pi^{(n)}(x, y) = \frac{d}{m(y)}.$$

If x is transient, then

$$\lim_{\substack{n \to \infty \\ n=r \bmod d}} \pi^{(n)}(x, y) = f(r, x, y) \frac{d}{m(y)}$$

where

$$f(r, x, y) = P_x\{X_{kd+r} = y \text{ for some } k \geq 0\}.$$

PROOF: The only statement that needs an explanation is the last one. The chain starting from a transient state x may at some time get into a positive recurrent equivalence class $\mathcal{X}_j$ with period d. If it does, it never leaves that class and so gets absorbed in that class. The probability of this is $f(x, y)$ where y can be any state in $\mathcal{X}_j$. However, if the period d is greater than 1, there will be cyclical subclasses $C_1, C_2, \ldots, C_d$ of $\mathcal{X}_j$. Depending on which subclass the chain enters and when, the phase of its future is determined. There are d such possible phases. For instance, if the subclasses are ordered in the correct way, getting into C_1 at time n is the same as getting into C_2 at time $n + 1$, and so on. $f(r, x, y)$ is the probability of getting into the equivalence class in a phase that visits the cyclical subclass containing y at times n that are equal to $r \mod d$. □

4.7. Some Examples

EXAMPLE 4.1 (Simple Random Walk). If $\mathcal{X} = \mathbb{Z}^d$, the integral lattice in $\mathbb{R}^d$, a random walk is a Markov chain with transition probability $\pi(x, y) = p(y - x)$ where $\{p(z)\}$ specifies the probability distribution of a single step. We will assume for simplicity that $p(z) = 0$ except when $z \in F$ where F consists of the $2d$ neighbors of 0 and $p(z) = \frac{1}{2d}$ for each $z \in F$. For $\xi \in \mathbb{R}^d$ the characteristic

4.7. SOME EXAMPLES

function $\hat{p}(\xi)$ of $p(\cdot)$ is given by $\frac{1}{d}(\cos\xi_1 + \cos\xi_2 + \cdots + \cos\xi_d)$. The chain is easily seen to be irreducible but periodic of period 2. Return to the starting point is possible only after an even number of steps,

$$\pi^{(2n)}(0,0) = \left(\frac{1}{2\pi}\right)^d \int_{\mathbb{T}^d} [\hat{p}(\xi)]^{2n}\, d\xi \simeq \frac{C}{n^{\frac{d}{2}}}.$$

To see this asymptotic behavior, let us first note that the integration can be restricted to the set where $|\hat{p}(\xi)| \geq 1 - \delta$ or near the two points $(0,0,\ldots,0)$ and $(\pi,\pi,\ldots,\pi)$ where $|\hat{p}(\xi)| = 1$. Since the behavior is similar at both points, let us concentrate near the origin.

$$\frac{1}{d}\sum_{j=1}^{d}\cos\xi_j \leq 1 - c\sum_j \xi_j^2 \leq \exp\left[-c\sum_j \xi_j^2\right] \quad \text{for some } c > 0$$

and

$$\left[\frac{1}{d}\sum_{j=1}^{d}\cos\xi_j\right]^{2n} \leq \exp\left[-2nc\sum_j \xi_j^2\right],$$

and with a change of variables the upper bound is clear. We have a similar lower bound as well. The random walk is recurrent if $d = 1$ or 2 but transient if $d \geq 3$.

EXERCISE 4.13. If the distribution $p(\cdot)$ is arbitrary, determine when the chain is irreducible and when it is irreducible and aperiodic.

EXERCISE 4.14. If $\sum_z zp(z) = m \neq 0$, conclude that the chain is transient by an application of the strong law of large numbers.

EXERCISE 4.15. If $\sum_z zp(z) = m = 0$ and if the covariance matrix given by $\sum_z z_i z_j p(z) = \sigma_{i,j}$ is nondegenerate, show that the transience or recurrence is determined by the dimension as in the case of the nearest-neighbor random walk.

EXERCISE 4.16. Can you make sense of the formal calculation

$$\sum_n \pi^{(n)}(0,0) = \sum_n \left(\frac{1}{2\pi}\right)^d \int_{\mathbb{T}^d}[\hat{p}(\xi)]^n\, d\xi$$

$$= \left(\frac{1}{2\pi}\right)^d \int_{\mathbb{T}^d}\frac{1}{(1-\hat{p}(\xi))}\, d\xi = \left(\frac{1}{2\pi}\right)^d \int_{\mathbb{T}^d}\operatorname{Re}\left[\frac{1}{1-\hat{p}(\xi)}\right] d\xi$$

to conclude that a necessary and sufficient condition for transience or recurrence is the convergence or divergence of the integral

$$\int_{\mathbb{T}^d}\operatorname{Re}\left[\frac{1}{1-\hat{p}(\xi)}\right] d\xi \quad \text{with an integrand} \quad \operatorname{Re}\left[\frac{1}{1-\hat{p}(\xi)}\right]$$

that is seen to be nonnegative?

Hint. Consider instead the sum

$$\sum_{n=0}^{\infty} \rho^n \pi^{(n)}(0,0) = \sum_n \left(\frac{1}{2\pi}\right)^d \int_{\mathbb{T}^d} \rho^n [\hat{p}(\xi)]^n \, d\xi$$

$$= \left(\frac{1}{2\pi}\right)^d \int_{\mathbb{T}^d} \frac{1}{(1-\rho\hat{p}(\xi))} \, d\xi = \left(\frac{1}{2\pi}\right)^d \int_{\mathbb{T}^d} \text{Re}\left[\frac{1}{1-\rho\hat{p}(\xi)}\right] d\xi$$

for $0 < \rho < 1$ and let $\rho \to 1$.

EXAMPLE 4.2 (The Queue Problem). In the example of customers arriving, except in the trivial cases of $p_0 = 0$ or $p_0 + p_1 = 1$, the chain is irreducible and aperiodic. Since the service rate is at most 1 if the arrival rate $m = \sum_j j \, p_j > 1$, then the queue will get longer and by an application of the law of large numbers it is seen that the queue length will become infinite as time progresses. This is the transient behavior of the queue. If $m < 1$, one can expect the situation to be stable and there should be an asymptotic distribution for the queue length. If $m = 1$, it is the borderline case, and one should probably expect this to be the null recurrent case. The actual proofs are not hard. In time n the actual number of customers served is at most n because the queue may sometimes be empty. If $\{\xi_i : i \geq 1\}$ are the number of new customers arriving at time i and X_0 is the initial number in the queue, then the number X_n in the queue at time n satisfies $X_n \geq X_0 + (\sum_{i=1}^n \xi_i) - n$ and if $m > 1$, it follows from the law of large numbers that $\lim_{n\to\infty} X_n = +\infty$, thereby establishing transience.

To prove positive recurrence when $m < 1$, it is sufficient to prove that the equation

$$\sum_x q(x) \pi(x, y) = q(y)$$

has a nontrivial nonnegative solution such that $\sum_x q(x) < \infty$. We shall proceed to show that this is indeed the case. Since the equation is linear, we can always normalize the solution so that $\sum_x q(x) = 1$. By iteration

$$\sum_x q(x) \pi^{(n)}(x, y) = q(y)$$

for every n. If $\lim_{n\to\infty} \pi^{(n)}(x, y) = 0$ for every x and y, because $\sum_x q(x) = 1 < \infty$, by the bounded convergence theorem the right-hand side tends to 0 as $n \to \infty$. Therefore $q \equiv 0$ and is trivial. This rules out the transient and the null recurrent case. In our case $\pi(0, y) = p_y$ and $\pi(x, y) = p_{y-x+1}$ if $y \geq x - 1$ and $x \geq 1$. In all other cases $\pi(x, y) = 0$. The equations for $\{q_x = q(x)\}$ are then

(4.13) $$q_0 p_y + \sum_{x=1}^{y+1} q_x p_{y-x+1} = q_y \quad \text{for } y \geq 0.$$

Multiplying equation (4.13) by z^y and summing from 0 to ∞, we get

$$q_0 P(z) + \frac{1}{z} P(z)[Q(z) - q_0] = Q(z)$$

4.7. SOME EXAMPLES

where $P(z)$ and $Q(z)$ are the generating functions

$$P(z) = \sum_{x=0}^{\infty} p_x z^x \quad \text{and} \quad Q(z) = \sum_{x=0}^{\infty} q_x z^x.$$

We can solve for Q to get

$$\frac{Q(z)}{q_0} = P(z)\left[1 - \frac{P(z)-1}{z-1}\right]^{-1}$$

$$= P(z) \sum_{k=0}^{\infty} \left[\frac{P(z)-1}{z-1}\right]^k = P(z) \sum_{k=0}^{\infty} \left[\sum_{j=1}^{\infty} p_j(1+z+\cdots+z^{j-1})\right]^k,$$

which is a power series in z with nonnegative coefficients. If $m < 1$, we can let $z \to 1$ to get

$$\frac{Q(1)}{q_0} = \sum_{k=0}^{\infty} \left[\sum_{j=1}^{\infty} j p_j\right]^k = \sum_{k=0}^{\infty} m^k = \frac{1}{1-m} < \infty$$

proving positive recurrence.

The case $m = 1$ is a little bit harder. The calculations carried out earlier are still valid and we know in this case that there exists $q(x) \geq 0$ such that $q(x) < \infty$ for each x, $\sum_x q(x) = \infty$, and

$$\sum_x q(x)\pi(x, y) = q(y).$$

In other words, the chain admits an infinite invariant measure. Such a chain cannot be positive recurrent. To see this we note

$$q(y) = \sum_x \pi^{(n)}(x, y) q(x),$$

and if the chain were positive recurrent

$$\lim_{n \to \infty} \pi^{(n)}(x, y) = \tilde{q}(y)$$

would exist and $\sum_y \tilde{q}(y) = 1$. By Fatou's lemma

$$q(y) \geq \sum_x \tilde{q}(y) q(x) = \infty,$$

giving us a contradiction. To decide between transience and null recurrence, a more detailed investigation is needed. We will outline a general procedure.

Suppose we have a state x_0 that is fixed and would like to calculate $F_{x_0}(\ell) = P_{x_0}\{\tau_{x_0} \leq \ell\}$. If we can do this, then we can answer questions about transience, recurrence, etc. If $\lim_{\ell \to \infty} F_{x_0}(\ell) < 1$, then the chain is transient and otherwise recurrent. In the recurrent case the convergence or divergence of

$$E^{P_{x_0}}\{\tau_{x_0}\} = \sum_{\ell}[1 - F_{x_0}(\ell)]$$

determines if it is positive or null recurrent. If we can determine

$$F_y(\ell) = P_y\{\tau_{x_0} \leq \ell\} \quad \text{for } y \neq x_0,$$

then

$$F_{x_0}(\ell) = \pi(x_0, x_0) + \sum_{y \neq x_0} \pi(x_0, y) F_y(\ell - 1) \quad \text{for } \ell \geq 1.$$

We shall outline a procedure for determining

$$U(\lambda, y) = E_y[\exp[-\lambda \tau_{x_0}]] \quad \text{for } \lambda > 0.$$

Clearly $U(x) = U(\lambda, x)$ satisfies

(4.14) $$U(x) = e^{-\lambda} \sum_y \pi(x, y) U(y) \quad \text{for } x \neq x_0 \text{ and } U(x_0) = 1.$$

One would hope that if we solve for these equations, then we have our U. This requires uniqueness. Since our U is bounded in fact, by 1, it is sufficient to prove uniqueness within the class of bounded solutions of equation (4.14). We will now establish that any bounded solution U of equation (4.14), with $U(x_0) = 1$, is given by

$$U(y) = U(\lambda, y) = E_y[\exp[-\lambda \tau_{x_0}]].$$

Let us define $E_n = \{X_1 \neq x_0, X_2 \neq x_0, \ldots, X_{n-1} \neq x_0, X_n = x_0\}$. Then we will prove, by induction, that for any solution U of equation (4.14), with $U(\lambda, x_0) = 1$,

(4.15) $$U(y) = \sum_{j=1}^n e^{-\lambda j} P_y\{E_j\} + e^{-\lambda n} \int_{\tau_{x_0} > n} U(X_n) dP_y.$$

By letting $n \to \infty$ we would obtain

$$U(y) = \sum_{j=1}^\infty e^{-\lambda j} P_y\{E_j\} = E^{P_y}\{e^{-\lambda \tau_{x_0}}\}$$

because U is bounded and $\lambda > 0$; then

$$\int_{\tau_{x_0} > n} U(X_n) dP_y = e^{-\lambda} \int_{\tau_{x_0} > n} \left[\sum_y \pi(X_n, y) U(y)\right] dP_y$$

$$= e^{-\lambda} P_y\{E_{n+1}\} + e^{-\lambda} \int_{\tau_{x_0} > n} \left[\sum_{y \neq x_0} \pi(X_n, y) U(y)\right] dP_y$$

$$= e^{-\lambda} P_y\{E_{n+1}\} + e^{-\lambda} \int_{\tau_{x_0} > n+1} U(X_{n+1}) dP_y,$$

completing the induction argument.

4.7. SOME EXAMPLES

In our case, if we take $x_0 = 0$ and try $U_\sigma(x) = e^{-\sigma x}$ with $\sigma > 0$, for $x \geq 1$

$$\sum_y \pi(x,y)U_\sigma(y) = \sum_{y \geq x-1} e^{-\sigma y} p_{y-x+1}$$

$$= \sum_{y \geq 0} e^{-\sigma(x+y-1)} p_y = e^{-\sigma x} e^\sigma \sum_{y \geq 0} e^{-\sigma y} p_y = \psi(\sigma) U_\sigma(x)$$

where

$$\psi(\sigma) = e^\sigma \sum_{y \geq 0} e^{-\sigma y} p_y.$$

Let us solve $e^\lambda = \psi(\sigma)$ for σ, which is the same as solving $\log \psi(\sigma) = \lambda$ for $\lambda > 0$ to get a solution $\sigma = \sigma(\lambda) > 0$. Then

$$U(\lambda, x) = e^{-\sigma(\lambda)x} = E^{P_x}\{e^{-\lambda \tau_0}\}.$$

We see now that recurrence is equivalent to $\sigma(\lambda) \to 0$ as $\lambda \downarrow 0$ and positive recurrence to $\sigma(\lambda)$ being differentiable at $\lambda = 0$. The function $\log \psi(\sigma)$ is convex and its slope at the origin is $1 - m$. If $m > 1$ it dips below 0 initially for $\sigma > 0$ and then comes back up to 0 for some positive σ_0 before turning positive for good. In that situation $\lim_{\lambda \downarrow 0} \sigma(\lambda) = \sigma_0 > 0$, and that is transience. If $m < 1$, then $\log \psi(\sigma)$ has a positive slope at the origin and $\sigma'(0) = \frac{1}{\psi'(0)} = \frac{1}{1-m} < \infty$. If $m = 1$, then $\log \psi$ has zero slope at the origin and $\sigma'(0) = \infty$. This concludes the discussion of this problem.

EXAMPLE 4.3 (The Urn Problem). We now turn to a discussion of the urn problem.

$$\pi(p, q; p+1, q) = \frac{p}{p+q} \quad \text{and} \quad \pi(p, q; p, q+1) = \frac{q}{p+q}$$

and π is zero otherwise. In this case the equation

$$F(p, q) = \frac{p}{p+q} F(p+1, q) + \frac{q}{p+q} F(p, q+1) \quad \text{for all } p, q,$$

which will play a role later, has lots of solutions. In particular, $F(p, q) = \frac{p}{p+q}$ is one, and for any $0 < x < 1$

$$F_x(p, q) = \frac{1}{\beta(p, q)} x^{p-1}(1-x)^{q-1} \quad \text{where} \quad \beta(p, q) = \frac{\Gamma(p)\Gamma(q)}{\Gamma(p+q)}$$

is a solution as well. The former is defined on $p + q > 0$, whereas the latter is defined only on $p > 0, q > 0$. Actually, if p or q is initially 0 it remains so forever, and there is nothing to study in that case. If f is a continuous function on $[0, 1]$, then

$$F_f(p, q) = \int_0^1 F_x(p, q) f(x) dx$$

is a solution, and if we want we can extend F_f by making it $f(1)$ on $q = 0$ and $f(0)$ on $p = 0$. It is a simple exercise to verify

$$\lim_{\substack{p,q \to \infty \\ \frac{p}{q} \to x}} F_f(p, q) = f(x)$$

for any continuous f on $[0, 1]$. We will show that the ratio $\xi_n = \frac{p_n}{p_n+q_n}$, which is random, stabilizes asymptotically (i.e., has a limit) to a random variable ξ, and if we start from p, q, the distribution of ξ is the beta distribution on $[0, 1]$ with density
$$F_x(p, q) = \frac{1}{\beta(p, q)} x^{p-1}(1-x)^{q-1}.$$

Suppose we have a Markov chain on some state space $\mathcal{X}$ with transition probability $\pi(x, y)$ and $U(x)$ is a bounded function on $\mathcal{X}$ that solves
$$U(x) = \sum_y \pi(x, y) U(y).$$

Such functions are called (*bounded*) *harmonic functions* for the chain. Consider the random variables $\xi_n = U(X_n)$ for such a harmonic function. ξ_n are uniformly bounded by the bound for U. If we let $\eta_n = \xi_n - \xi_{n-1}$, an elementary calculation reveals
$$E^{P_x}\{\eta_{n+1}\} = E^{P_x}\{U(X_{n+1}) - U(X_n)\} = E^{P_x}\{E^{P_x}[\{U(X_{n+1}) - U(X_n)\} \mid \mathcal{F}_n]\}$$
where $\mathcal{F}_n$ is the σ-field generated by $X_0, X_1, \ldots, X_n$. But
$$E^{P_x}[\{U(X_{n+1}) - U(X_n)\} \mid \mathcal{F}_n] = \sum_y \pi(X_n, y)[U(y) - U(X_n)] = 0.$$

A similar calculation shows that
$$E^{P_x}\{\eta_n \eta_m\} = 0$$
for $m \neq n$. If we write
$$U(X_n) = U(X_0) + \eta_1 + \eta_2 + \cdots + \eta_n,$$
this is an orthogonal sum in $L_2[P_x]$, and because U is bounded
$$E^{P_x}\{|U(X_n)|^2\} = |U(x)|^2 + \sum_{i=1}^n E^{P_x}\{|\eta_i|^2\} \leq C$$
is bounded in n. Therefore, $\lim_{n \to \infty} U(X_n) = \xi$ exists in $L_2[P_x]$ and $E^{P_x}\{\xi\} = U(x)$. Actually, the limit exists almost surely, and we will show it when we discuss martingales later. In our example, if we take $U(p, q) = \frac{p}{p+q}$, as remarked earlier, this is a harmonic function bounded by 1, and therefore
$$\lim_{n \to \infty} \frac{p_n}{p_n + q_n} = \xi$$
exists in $L_2[P_x]$. Moreover, if we take $U(p, q) = F_f(p, q)$ for some continuous f on $[0, 1]$, because $F_f(p, q) \to f(x)$ as $p, q \to \infty$ and $\frac{p}{q} \to x$, $U(p_n, q_n)$ has a limit as $n \to \infty$ and this limit has to be $f(\xi)$. On the other hand,
$$E^{P_{p,q}}\{U(p_n, q_n)\} = U(p_0, q_0) = F_f(p_0, q_0)$$
$$= \frac{1}{\beta(p_0, q_0)} \int_0^1 f(x) x^{p_0-1}(1-x)^{q_0-1} dx,$$

giving us
$$E^{P_{p,q}}\{f(\xi)\} = \frac{1}{\beta(p,q)} \int_0^1 f(x) x^{p-1}(1-x)^{q-1} dx,$$
thereby identifying the distribution of ξ under $P_{p,q}$ as the beta distribution with the right parameters.

EXAMPLE 4.4 (Branching Process). Consider a population in which each individual member replaces itself at the beginning of each day by a random number of offsprings. Every individual has the same probability distribution, but the number of offsprings for different individuals are distributed independently of each other. The distribution of the number N of offsprings of a single individual is given by $P[N = i] = p_i$ for $i \geq 0$. If there are X_n individuals in the population on a given day, then the number of individuals X_{n+1} present on the next day has the representation
$$X_{n+1} = N_1 + N_2 + \cdots + N_{X_n}$$
as the sum of X_n independent random variables, each having the offsprings distribution $\{p_i : i \geq 0\}$. X_n is seen to be a Markov chain on the set of nonnegative integers. Note that if X_n ever becomes zero, i.e., if every member on a given day produces no offspring, then the population remains extinct.

If one uses generating functions, then the transition probability $\pi_{i,j}$ of the chain are
$$\sum_j \pi_{i,j} z^j = \left[\sum_j p_j z^j\right]^i.$$
What is the long-time behavior of the chain?

Let us denote by m the expected number of offsprings of any individual, i.e.,
$$m = \sum_{i \geq 0} i p_i.$$
Then
$$E[X_{n+1} \mid \mathcal{F}_n] = m X_n.$$

(1) If $m < 1$, then the population becomes extinct sooner or later. This is easy to see. Consider
$$E\left[\sum_{n \geq 0} X_n \mid \mathcal{F}_0\right] = \sum_{n \geq 0} m^n X_0 = \frac{1}{1-m} X_0 < \infty.$$
By an application of Fubini's theorem, if $S = \sum_{n \geq 0} X_n$, then
$$E[S \mid X_0 = i] = \frac{i}{1-m} < \infty,$$
proving that $P[S < \infty] = 1$. In particular,
$$P\left[\lim_{n \to \infty} X_n = 0\right] = 1.$$

(2) If $m = 1$ and $p_1 = 1$, then $X_n \equiv X_0$ and the population size never changes, each individual replacing itself every time by exactly one offspring.

(3) If $m = 1$ and $p_1 < 1$, then $p_0 > 0$, and there is a positive probability $q(i) = q^i$ that the population becomes extinct when it starts with i individuals. Here q is the probability of the population becoming extinct when we start with $X_0 = 1$. If we initially have i individuals, each of the i family lines have to become extinct for the entire population to become extinct. The number q must therefore be a solution of the equation

$$q = P(q)$$

where $P(z)$ is the generating function

$$P(z) = \sum_{i \geq 0} p_i z^i.$$

If we show that the equation $P(z) = z$ has only the solution $z = 1$ in $0 \leq z \leq 1$, then the population becomes extinct with probability 1, although $E[S] = \infty$ in this case. If $P(1) = 1$ and $P(a) = a$ for some $0 \leq a < 1$, then by the mean-value theorem applied to $P(z) - z$, we must have $P'(z) = 1$ for some $0 < z < 1$. But if $0 < z < 1$

$$P'(z) = \sum_{i \geq 1} i z^{i-1} p_i < \sum_{i \geq 1} i p_i = 1,$$

a contradiction.

(4) If $m > 1$ but $p_0 = 0$, the problem is trivial. There is no chance of the population becoming extinct. Let us assume that $p_0 > 0$. The equation $P(z) = z$ has another solution $z = q$ besides $z = 1$ in the range $0 < z < 1$. This is seen by considering the function $g(z) = P(z) - z$. We have $g(0) > 0, g(1) = 0, g'(1) > 0$, which implies another root. But $g(z)$ is convex, and therefore there can be at most one other root. If we can rule out the possibility of the extinction probability being equal to 1, then this root q must be the extinction probability when we start with a single individual at time 0. Let us denote by q_n the probability of extinction within n days. Then

$$q_{n+1} = \sum_i p_i q_n^i = P(q_n)$$

and $q_1 < 1$. A simple consequence of the monotonicity of $P(z)$ and the inequalities $P(z) > z$ for $z < q$ and $P(z) < z$ for $z > q$ is that if we start with any $a < 1$ and iterate $q_{n+1} = P(q_n)$ with $q_1 = a$, then $q_n \to q$.

If the population does not become extinct, one can show that it has to grow indefinitely. This is best done using martingales, and we will revisit this example later as Example 5.6.

EXAMPLE 4.5. Let $\mathcal{X}$ be the set of integers. Assume that transitions from x are possible only to $x - 1, x$, and $x + 1$. The transition matrix $\pi(x, y)$ appears as

4.7. SOME EXAMPLES

a tridiagonal matrix with $\pi(x, y) = 0$ unless $|x - y| \leq 1$. For simplicity, let us assume that $\pi(x, x)$, $\pi(x, x-1)$, and $\pi(x, x+1)$ are positive for all x.

The chain is then irreducible and aperiodic. Let us try to solve for

$$U(x) = P_x\{\tau_0 = \infty\},$$

which satisfies the equation

$$U(x) = \pi(x, x-1)U(x-1) + \pi(x, x)U(x) + \pi(x, x+1)U(x+1)$$

for $x \neq 0$ with $U(0) = 0$. The equations decouple into a set for $x > 0$ and a set for $x < 0$. If we denote by $V(x) = U(x+1) - U(x)$ for $x \geq 0$, then we always have

$$U(x) = \pi(x, x-1)U(x) + \pi(x, x)U(x) + \pi(x, x+1)U(x)$$

so that

$$\pi(x, x-1)V(x-1) - \pi(x, x+1)V(x) = 0 \quad \text{or} \quad \frac{V(x)}{V(x-1)} = \frac{\pi(x, x-1)}{\pi(x, x+1)}$$

and therefore

$$V(x) = V(0)\prod_{i=1}^{x} \frac{\pi(i, i-1)}{\pi(i, i+1)} \quad \text{and} \quad U(x) = V(0)\left[\sum_{y=1}^{x-1}\prod_{i=1}^{y} \frac{\pi(i, i-1)}{\pi(i, i+1)}\right].$$

If the chain is to be transient, we must have for some choice of $V(0)$, $0 \leq U(x) \leq 1$ for all $x > 0$, and this will be possible only if

$$\sum_{y=1}^{\infty}\prod_{i=1}^{y} \frac{\pi(i, i-1)}{\pi(i, i+1)} < \infty,$$

which then becomes a necessary condition for

$$P_x\{\tau_0 = \infty\} > 0$$

for $x > 0$. There is a similar condition on the negative side,

$$\sum_{y=1}^{\infty}\prod_{i=1}^{y} \frac{\pi(-i, -i+1)}{\pi(-i, -i-1)} < \infty.$$

Transience needs at least one of the two series to converge. Actually, the converse is also true. If, for instance, the series on the positive side converges, then we get a function $U(x)$ with $0 \leq U(x) \leq 1$ and $U(0) = 0$ that satisfies

$$U(x) = \pi(x, x-1)U(x-1) + \pi(x, x)U(x) + \pi(x, x+1)U(x+1),$$

and by iteration one can prove that for each n,

$$U(x) = \int_{\tau_0 > n} U(X_n)dP_x \leq P\{\tau_0 > n\},$$

so the existence of a nontrivial U implies transience.

EXERCISE 4.17. Determine the conditions for positive recurrence in the previous example.

EXERCISE 4.18. We replace the set of integers by the set of nonnegative integers and assume that $\pi(0, y) = 0$ for $y \geq 2$. Such processes are called *birth-and-death processes*. Work out the conditions in that case.

EXERCISE 4.19. In the special case where we have a birth-and-death process with $\pi(0, 1) = \pi(0, 0) = \frac{1}{2}$ and for $x \geq 1$, $\pi(x, x) = \frac{1}{3}$, $\pi(x, x - 1) = \frac{1}{3} + a_x$, and $\pi(x, x + 1) = \frac{1}{3} - a_x$ with $a_x = \frac{\lambda}{x^\alpha}$ for large x, find conditions on positive α and real λ for the chain to be transient, null recurrent, and positive recurrent.

EXERCISE 4.20. The notion of a Markov chain makes sense for a finite chain $X_0, X_1, \ldots, X_n$. Formulate it precisely. Show that if the chain $\{X_j : 0 \leq j \leq n\}$ is Markov, so is the reversed chain $\{Y_j : 0 \leq j \leq n\}$ where $Y_j = X_{n-j}$ for $0 \leq j \leq n$. Can the transition probabilities of the reversed chain be determined by the transition probabilities of the forward chain? If the forward chain has stationary transition probabilities, does the same hold true for the reversed chain? What if we assume that the chain has an invariant probability distribution and we initialize the chain to start with an initial distribution that is the invariant distribution?

EXERCISE 4.21. Consider the simple chain on nonnegative integers with the following transition probabilities: $\pi(0, x) = p_x$ for $x \geq 0$ with $\sum_{x=0}^{\infty} p_x = 1$. For $x > 0$, $\pi(x, x - 1) = 1$ and $\pi(x, y) = 0$ for all other y. Determine conditions on $\{p_x\}$ in order that the chain may be transient, null recurrent, or positive recurrent. Determine the invariant probability measure in the positive recurrent case.

EXERCISE 4.22. Show that any null recurrent equivalence class must necessarily contain an infinite number of states. In particular, any Markov chain with a finite state space has only transient and positive recurrent states and, moreover, the set of positive recurrent states must be nonempty.

CHAPTER 5

Martingales

5.1. Definitions and Properties

The theory of martingales plays a very important and useful role in the study of stochastic processes. A formal definition is given below.

DEFINITION 5.1 Let $(\Omega, \mathcal{F}, P)$ be a probability space. A *martingale* sequence of length n is a chain $X_1, X_2, \ldots, X_n$ of random variables and corresponding sub σ-fields $\mathcal{F}_1, \mathcal{F}_2, \ldots, \mathcal{F}_n$ that satisfy the following relations:
 (1) Each X_i is an integrable random variable which is measurable with respect to the corresponding σ-field $\mathcal{F}_i$.
 (2) The σ-fields $\mathcal{F}_i$ are increasing, i.e., $\mathcal{F}_i \subset \mathcal{F}_{i+1}$ for every i.
 (3) For every $i \in [1, 2, \ldots, n-1]$, we have the relation
$$X_i = E\{X_{i+1} \mid \mathcal{F}_i\} \quad \text{a.e. } P.$$

REMARK 5.1. We can have an infinite martingale sequence $\{(X_i, \mathcal{F}_i) : i \geq 1\}$ which requires only that for every n, $\{(X_i, \mathcal{F}_i) : 1 \leq i \leq n\}$ be a martingale sequence of length n. This is the same as conditions (1), (2), and (3) above except that they have to be true for every $i \geq 1$.

REMARK 5.2. From the properties of conditional expectations we see that $E\{X_i\} = E\{X_{i+1}\}$ for every i, and therefore $E\{X_i\} = c$ for some c. We can define $\mathcal{F}_0$ to be the trivial σ-field consisting of $\{\varnothing, \Omega\}$ and $X_0 = c$. Then $\{(X_i, \mathcal{F}_i) : i \geq 0\}$ is a martingale sequence as well.

REMARK 5.3. We can define $Y_{i+1} = X_{i+1} - X_i$ so that $X_i = c + \sum_{1 \leq j \leq i} Y_j$ and property (3) reduces to
$$E\{Y_{i+1} \mid \mathcal{F}_i\} = 0 \quad \text{a.e. } P.$$

Such sequences are called *martingale differences*. If Y_i is a sequence of independent random variables with mean 0, for each i, we can take $\mathcal{F}_i$ to be the σ-field generated by the random variables $\{Y_j : 1 \leq j \leq i\}$, and $X_i = c + \sum_{1 \leq j \leq i} Y_j$ will be a martingale relative to the σ-fields $\mathcal{F}_i$.

REMARK 5.4. We can generate martingale sequences by the following procedure: Given any increasing family of σ-fields $\{\mathcal{F}_j\}$ and any integrable random variable X on $(\Omega, \mathcal{F}, P)$, we take $X_i = E\{X \mid \mathcal{F}_i\}$ and it is easy to check that $\{(X_i, \mathcal{F}_i)\}$ is a martingale sequence. Of course, every finite martingale sequence is generated this way, for we can always take X to be X_n, the last one. For infinite sequences this raises an important question that we will answer later.

If one participates in a "fair" gambling game, the asset X_n of the player at time n is supposed to be a martingale. One can take for $\mathcal{F}_n$ the σ-field of all the results of the game through time n. The condition $E[X_{n+1} - X_n \mid \mathcal{F}_n] = 0$ is the assertion that the game is neutral irrespective of past history.

A related notion is that of a super- or submartingale. If, in the definition of a martingale, we replace the equality in (3) by an inequality, we get super- or submartingales.

For a *submartingale* we demand the relation

(3a) for every i,
$$X_i \leq E\{X_{i+1} \mid \mathcal{F}_i\} \quad \text{a.e. } P,$$

while for a *supermartingale* the relation is

(3b) for every i,
$$X_i \geq E\{X_{i+1} \mid \mathcal{F}_i\} \quad \text{a.e. } P.$$

LEMMA 5.1 *If $\{(X_i, \mathcal{F}_i)\}$ is a martingale and φ is a convex (or concave) function of one variable such that $\varphi(X_n)$ is integrable for every n, then $\{(\varphi(X_n), \mathcal{F}_n)\}$ is a sub- (or super-) martingale.*

PROOF: An easy consequence of Jensen's inequality (4.2) for conditional expectations. □

REMARK 5.5. A particular case is $\phi(x) = |x|^p$ with $1 \leq p < \infty$. For any martingale $(X_n, \mathcal{F}_n)$ and $1 \leq p < \infty$, $(|X_n|^p, \mathcal{F}_n)$ is a submartingale provided $E[|X_n|^p] < \infty$.

THEOREM 5.2 (Doob's Inequality) *Suppose $\{X_j\}$ is a martingale sequence of length n. Then*

(5.1) $$P\left\{\omega : \sup_{1 \leq j \leq n} |X_j| \geq \ell\right\} \leq \frac{1}{\ell} \int_{\{\sup_{1 \leq j \leq n} |X_j| \geq \ell\}} |X_n| dP \leq \frac{1}{\ell} \int |X_n| dP.$$

PROOF: Let us define $S(\omega) = \sup_{1 \leq j \leq n} |X_j(\omega)|$. Then
$$\{\omega : S(\omega) \geq \ell\} = E = \bigcup_j E_j$$

is written as a disjoint union, where
$$E_j = \{\omega : |X_1(\omega)| < \ell, |X_2(\omega)| < \ell, \ldots, |X_{j-1}(\omega)| < \ell, |X_j(\omega)| \geq \ell\}.$$

We have

(5.2) $$P(E_j) \leq \frac{1}{\ell} \int_{E_j} |X_j| dP \leq \frac{1}{\ell} \int_{E_j} |X_n| dP.$$

The second inequality in (5.2) follows from the fact that $|x|$ is a convex function of x, and therefore $|X_j|$ is a submartingale. In particular, $E\{|X_n| \mid \mathcal{F}_j\} \geq |X_j|$ a.e. P and $E_j \in \mathcal{F}_j$. Summing up (5.2) over $j = 1, 2, \ldots, n$ we obtain the theorem. □

5.1. DEFINITIONS AND PROPERTIES

REMARK 5.6. We could have started with

$$P(E_j) \leq \frac{1}{\ell^p} \int_{E_j} |X_j|^p \, dP$$

and obtained for $p \geq 1$

(5.3) $$P(E) \leq \frac{1}{\ell^p} \int_E |X_n|^p \, dP.$$

Compare it with (3.9) for $p = 2$.

This simple inequality has various implications. For example,

COROLLARY 5.3 (Doob's Inequality) *Let $\{X_j : 1 \leq j \leq n\}$ be a martingale. Then if, as before, $S(\omega) = \sup_{1 \leq j \leq n} |X_j(\omega)|$, we have*

$$E[S^p] \leq \left(\frac{p}{p-1}\right)^p E[|X_n|^p].$$

The proof is a consequence of the following fairly general lemma:

LEMMA 5.4 *Suppose X and Y are two nonnegative random variables on a probability space such that*

$$P\{Y \geq \ell\} \leq \frac{1}{\ell} \int_{Y \geq \ell} X \, dP \quad \text{for every } \ell \geq 0;$$

then

$$\int Y^p \, dP \leq \left(\frac{p}{p-1}\right)^p \int X^p \, dP \quad \text{for every } p > 1.$$

PROOF: Let us denote the tail probability by $T(\ell) = P\{Y \geq \ell\}$. Then, with $\frac{1}{p} + \frac{1}{q} = 1$, i.e., $(p-1)q = p$,

$$\int Y^p \, dP = -\int_0^\infty y^p \, dT(y) = p \int_0^\infty y^{p-1} T(y) \, dy \quad \text{(integrating by parts)}$$

$$\leq p \int_0^\infty y^{p-1} \frac{dy}{y} \int_{Y \geq y} X \, dP \quad \text{(by assumption)}$$

$$= p \int X \left[\int_0^Y y^{p-2} \, dy\right] dP \quad \text{(by Fubini's theorem)}$$

$$= \frac{p}{p-1} \int X Y^{p-1} \, dP$$

$$\leq \frac{p}{p-1} \left[\int X^p \, dP\right]^{\frac{1}{p}} \left[\int Y^{q(p-1)} \, dP\right]^{\frac{1}{q}} \quad \text{(by Hölder's inequality)}$$

$$\leq \frac{p}{p-1} \left[\int X^p \, dP\right]^{\frac{1}{p}} \left[\int Y^p \, dP\right]^{1-\frac{1}{p}}.$$

This simplifies to

$$\int Y^p \, dP \leq \left(\frac{p}{p-1}\right)^p \int X^p \, dP$$

provided $\int Y^p \, dP$ is finite. In general, given Y, we can truncate it at level N to get $Y_N = \min(Y, N)$ and

$$P\{Y_N \geq \ell\} \leq P\{Y \geq \ell\} \leq \frac{1}{\ell} \int_{Y \geq \ell} X \, dP = \frac{1}{\ell} \int_{Y_N \geq \ell} X \, dP \quad \text{for } 0 < \ell \leq N$$

with $P\{Y_N \geq \ell\} = 0$ for $\ell > N$. This gives us uniform bounds on $\int Y_N^p \, dP$ and we can pass to the limit. So we have the strong implication that the finiteness of $\int X^p \, dP$ implies the finiteness of $\int Y^p \, dP$. □

EXERCISE 5.1. The analogue of Corollary 5.3 is false for $p = 1$. Construct a nonnegative martingale X_n with $E[X_n] \equiv 1$ such that $\xi = \sup_n X_n$ is not integrable. Consider $\Omega = [0, 1]$, $\mathcal{F}$ the Borel σ-field, and P the Lebesgue measure. Suppose we take $\mathcal{F}_n$ to be the σ-field generated by intervals with endpoints of the form $\frac{j}{2^n}$ for some integer j. It corresponds to a partition with 2^n sets. Consider the random variables

$$X_n(x) = \begin{cases} 2^n & \text{for } 0 \leq x \leq 2^{-n} \\ 0 & \text{for } 2^{-n} < x \leq 1. \end{cases}$$

Check that it is a martingale and calculate $\int \xi(x) dx$. This is the "winning" strategy of doubling one's bets until the losses are recouped.

EXERCISE 5.2. If X_n is a martingale such that the differences $Y_n = X_n - X_{n-1}$ are all square-integrable, show that for $n \neq m$, $E[Y_n Y_m] = 0$. Therefore,

$$E[X_n^2] = E[X_0^2] + \sum_{j=1}^{n} E[Y_j^2].$$

If, in addition, $\sup_n E[X_n^2] < \infty$, then show that there is a random variable X such that

$$\lim_{n \to \infty} E[|X_n - X|^2] = 0.$$

5.2. Martingale Convergence Theorems

If $\mathcal{F}_n$ is an increasing family of σ-fields and X_n is a martingale sequence with respect to $\mathcal{F}_n$, one can always assume without loss of generality that the full σ-field $\mathcal{F}$ is the smallest σ-field generated by $\bigcup_n \mathcal{F}_n$. If for some $p \geq 1$, $X \in L_p$ and we define $X_n = E[X \mid \mathcal{F}_n]$, then X_n is a martingale and by Jensen's inequality, $\sup_n E[|X_n|^p] \leq E[|X|^p]$. We would like to prove

THEOREM 5.5 *For $p \geq 1$, if $X \in L_p$, then $\lim_{n \to \infty} \|X_n - X\|_p = 0$.*

5.2. MARTINGALE CONVERGENCE THEOREMS

PROOF: Assume that X is a bounded function. Then by the properties of conditional expectation, $\sup_n \sup_\omega |X_n| < \infty$. In particular, $E[X_n^2]$ is uniformly bounded. By Exercise 5.2, $\lim_{n \to \infty} X_n = Y$ exists in L_2. By the properties of conditional expectations, for $A \in \mathcal{F}_m$,

$$\int_A Y\, dP = \lim_{n \to \infty} \int_A X_n\, dP = \int_A X\, dP.$$

This is true for all $A \in \mathcal{F}_m$ for any m. Since $\mathcal{F}$ is generated by $\bigcup_m \mathcal{F}_m$, the above relation is true for $A \in \mathcal{F}$. As X and Y are $\mathcal{F}$-measurable, we conclude that $X = Y$ a.e. P; see Exercise 4.1. For a sequence of functions that are bounded uniformly in n and ω, convergences in L_p are all equivalent and therefore convergence in L_2 implies the convergence in L_p for any $1 \le p < \infty$. If now $X \in L_p$ for some $1 \le p < \infty$, we can approximate it by $X' \in L_\infty$ so that $\|X' - X\|_p < \varepsilon$. Let us denote by X_n' the conditional expectations $E[X' \mid \mathcal{F}_n]$. By the properties of conditional expectations $\|X_n' - X_n\|_p \le \varepsilon$ for all n, and, as we saw earlier, $\|X_n' - X'\|_p \to 0$ as $n \to \infty$. It now follows that

$$\limsup_{n \to \infty} \|X_n - X\|_p \le 2\varepsilon,$$

and because $\varepsilon > 0$ is arbitrary we are done. $\square$

In general, if we have a martingale $\{X_n\}$, we wish to know when it comes from a random variable $X \in L_p$ in the sense that $X_n = E[X \mid \mathcal{F}_n]$.

THEOREM 5.6 *If, for some $p > 1$, a martingale $\{X_n\}$ is bounded in L_p in the sense that $\sup_n \|X_n\|_p < \infty$, then there is a random variable $X \in L_p$ such that $X_n = E[X \mid \mathcal{F}_n]$ for $n \ge 1$. In particular, $\|X_n - X\|_p \to 0$ as $n \to \infty$.*

PROOF: Suppose $\|X_n\|_p$ is uniformly bounded. For $p > 1$, since L_p is the dual of L_q with $\frac{1}{p} + \frac{1}{q} = 1$, bounded sets are weakly compact; see [3] or [7]. We can therefore choose a subsequence X_{n_j} that converges weakly in L_p to a limit. We call this limit X. Then consider $A \in \mathcal{F}_n$ for some fixed n. The function $\mathbf{1}_A(\cdot) \in L_q$.

$$\int_A X\, dP = \langle \mathbf{1}_A, X \rangle = \lim_{j \to \infty} \langle \mathbf{1}_A, X_{n_j} \rangle = \lim_{j \to \infty} \int_A X_{n_j}\, dP = \int_A X_n\, dP.$$

The last equality follows from the fact that $\{X_n\}$ is a martingale, $A \in \mathcal{F}_n$, and $n_j > n$ eventually. It now follows that $X_n = E[X \mid \mathcal{F}_n]$ and we can apply the preceding theorem. $\square$

EXERCISE 5.3. For $p = 1$ the result is false. Example 5.1 gives us at the same time a counterexample of an L_1-bounded martingale that does not converge in L_1 and so cannot be represented as $X_n = E[X \mid \mathcal{F}_n]$.

We can show that the convergence in the preceding theorems is also valid almost everywhere.

THEOREM 5.7 *Let $X \in L_p$ for some $p \ge 1$. Then the martingale $X_n = E[X \mid \mathcal{F}_n]$ converges to X for almost all ω with respect to P.*

PROOF: From Hölder's inequality $\|X\|_1 \leq \|X\|_p$. Clearly, it is sufficient to prove the theorem for $p = 1$. Let us denote by $\mathcal{M} \subset L_1$ the set of functions $X \in L_1$ for which the theorem is true. Clearly $\mathcal{M}$ is a linear subset of L_1. We will prove that it is closed in L_1 and that it is dense in L_1. If we denote by $\mathcal{M}_n$ the space of $\mathcal{F}_n$-measurable functions in L_1, then $\mathcal{M}_n$ is a closed subspace of L_1. By standard approximation theorems, $\bigcup_n \mathcal{M}_n$ is dense in L_1. Since it is obvious that $\mathcal{M} \supset \mathcal{M}_n$ for every n, it follows that $\mathcal{M}$ is dense in L_1. Let $Y_j \in \mathcal{M} \subset L_1$ and $Y_j \to X$ in L_1. Let us define $Y_{n,j} = E[Y_j \mid \mathcal{F}_n]$. With $X_n = E[X \mid \mathcal{F}_n]$, by Doob's inequality (5.1) and Jensen's inequality (4.2),

$$P\left\{\sup_{1 \leq n \leq N} |X_n| \geq \ell\right\} \leq \frac{1}{\ell} \int_{\{\omega: \sup_{1 \leq n \leq N} |X_n| \geq \ell\}} |X_N| dP \leq \frac{1}{\ell} E[|X_N|] \leq \frac{1}{\ell} E[|X|],$$

and therefore X_n is almost surely a bounded sequence. Since we know that $X_n \to X$ in L_1, it suffices to show that

$$\limsup_n X_n - \liminf_n X_n = 0 \quad \text{a.e. } P.$$

If we write $X = Y_j + (X - Y_j)$, then $X_n = Y_{n,j} + (X_n - Y_{n,j})$, and

$$\limsup_n X_n - \liminf_n X_n \leq \left[\limsup_n Y_{n,j} - \liminf_n Y_{n,j}\right]$$
$$+ \left[\limsup_n (X_n - Y_{n,j}) - \liminf_n (X_n - Y_{n,j})\right]$$
$$= \limsup_n (X_n - Y_{n,j}) - \liminf_n (X_n - Y_{n,j})$$
$$\leq 2 \sup_n |X_n - Y_{n,j}|.$$

Here we have used the fact that $Y_j \in \mathcal{M}$ for every j, and hence

$$\limsup_n Y_{n,j} - \liminf_n Y_{n,j} = 0 \quad \text{a.e. } P.$$

Finally,

$$P\left\{\limsup_n X_n - \liminf_n X_n \geq \varepsilon\right\} \leq P\left\{\sup_n |X_n - Y_{n,j}| \geq \frac{\varepsilon}{2}\right\}$$
$$\leq \frac{2}{\varepsilon} E[|X - Y_j|] = 0,$$

since the left-hand side is independent of j and the term on the right on the second line tends to 0 as $j \to \infty$. □

The only case where the situation is unclear is when $p = 1$. If X_n is an L_1-bounded martingale, it is not clear that it comes from an X. If it did arise from an X, then X_n would converge to it in L_1 and, in particular, would have to be uniformly integrable. The converse is also true.

THEOREM 5.8 *If X_n is a uniformly integrable martingale, then there is a random variable X such that $X_n = E[X \mid \mathcal{F}_n]$, and then, of course, $X_n \to X$ in L_1.*

PROOF: The uniform integrability of X_n implies the weak compactness in L_1, and if X is any weak limit of X_n (see [7]), it is not difficult to show, as in Theorem 5.5, that $X_n = E[X \mid \mathcal{F}_n]$. □

REMARK 5.7. Note that for $p > 1$, a martingale X_n that is bounded in L_p is uniformly integrable in L_p, i.e., $|X_n|^p$ is uniformly integrable. This is false for $p = 1$. The L_1-bounded martingale that we constructed earlier in Exercise 5.1 as a counterexample is not convergent in L_1 and therefore cannot be uniformly integrable. We will defer the analysis of L_1-bounded martingales to the next section.

5.3. Doob Decomposition Theorem

The simplest example of a submartingale is a sequence of functions that is nondecreasing in n for every (almost all) ω. In some sense, the simplest example is also the most general. More precisely, the decomposition theorem of Doob asserts the following:

THEOREM 5.9 (Doob Decomposition Theorem) *If $\{X_n : n \geq 1\}$ is a submartingale on $(\Omega, \mathcal{F}_n, P)$, then X_n can be written as $X_n = Y_n + A_n$, with the following properties*:

  (i) $(Y_n, \mathcal{F}_n)$ is a martingale.
  (ii) $A_{n+1} \geq A_n$ for almost all ω and for every $n \geq 1$.
  (iii) $A_1 \equiv 0$.
  (iv) For every $n \geq 2$, A_n is $\mathcal{F}_{n-1}$-measurable.

X_n determines Y_n and A_n uniquely.

PROOF: Let X_n be any sequence of integrable functions such that X_n is $\mathcal{F}_n$-measurable and is represented as $X_n = Y_n + A_n$, with Y_n and A_n satisfying (i), (iii), and (iv). Then

(5.4) $$A_n - A_{n-1} = E[X_n - X_{n-1} \mid \mathcal{F}_{n-1}]$$

are uniquely determined. Since $A_1 = 0$, all the A_n are uniquely determined as well. Property (ii) is then plainly equivalent to the submartingale property. To establish the representation, we define A_n inductively by (5.4). It is routine to verify that $Y_n = X_n - A_n$ is a martingale and the monotonicity of A_n is a consequence of the submartingale property. □

REMARK 5.8. Actually, given any sequence of integrable functions $\{X_n : n \geq 1\}$ such that X_n is $\mathcal{F}_n$-measurable, equation (5.4) along with $A_1 = 0$ defines $\mathcal{F}_{n-1}$-measurable functions that are integrable such that $X_n = Y_n + A_n$ and $(Y_n, \mathcal{F}_n)$ is a martingale. The decomposition is always unique. It is easy to verify from (5.4) that $\{A_n\}$ is increasing (or decreasing) if and only if $\{X_n\}$ is a sub- (or super-) martingale. Such a decomposition is called the *semimartingale decomposition*.

REMARK 5.9. It is the demand that A_n be $\mathcal{F}_{n-1}$-measurable that leads to uniqueness. If we have to deal with continuous time, this will become a thorny issue.

We now return to the study of L_1-bounded martingales. A nonnegative martingale is clearly L_1-bounded because $E[|X_n|] = E[X_n] = E[X_1]$. One easy way to generate L_1-bounded martingales is to take the difference of two nonnegative martingales. We have the converse as well.

THEOREM 5.10 *Let X_n be an L_1-bounded martingale. There are two nonnegative martingales Y_n and Z_n relative to the same σ-fields $\mathcal{F}_n$ such that $X_n = Y_n - Z_n$.*

PROOF: For each j and $n \geq j$, we define
$$Y_{j,n} = E[|X_n| \mid \mathcal{F}_j].$$
By the submartingale property of $|X_n|$,
$$Y_{j,n+1} - Y_{j,n} = E[(|X_{n+1}| - |X_n|) \mid \mathcal{F}_j] = E[E[(|X_{n+1}| - |X_n|) \mid \mathcal{F}_n] \mid \mathcal{F}_j] \geq 0$$
almost surely. $Y_{j,n}$ is nonnegative and $E[Y_{j,n}] = E[|X_n|]$ is bounded in n. By the monotone convergence theorem, for each j, there exists some Y_j in L_1 such that $Y_{j,n} \to Y_j$ in L_1 as $n \to \infty$. Since limits of martingales are again martingales, and $Y_{n,j}$ is a martingale for $n \geq j$, it follows that Y_j is a martingale. Moreover,
$$Y_j + X_j = \lim_{n \to \infty} E[|X_n| + X_n \mid \mathcal{F}_j] \geq 0 \quad \text{and} \quad X_j = (Y_j + X_j) - Y_j$$
does it! $\square$

We can assume that our nonnegative martingale has its expectation equal to 1 because we can always multiply by a suitable constant. Here is a way in which such martingales arise. Suppose we have a probability space $(\Omega, \mathcal{F}, P)$ and an increasing family of sub σ-fields $\mathcal{F}_n$ of $\mathcal{F}$ that generate $\mathcal{F}$. Suppose Q is another probability measure on $(\Omega, \mathcal{F})$ that may or may not be absolutely continuous with respect to P on $\mathcal{F}$. Let us suppose, however, that $Q \ll P$ on each $\mathcal{F}_n$, i.e., whenever $A \in \mathcal{F}_n$ and $P(A) = 0$, it follows that $Q(A) = 0$. Then the sequence of Radon-Nikodym derivatives
$$X_n = \left.\frac{dQ}{dP}\right|_{\mathcal{F}_n}$$
of Q with respect to P on $\mathcal{F}_n$ is a nonnegative martingale with expectation 1. It comes from an X if and only if $Q \ll P$ on $\mathcal{F}$, and this is the uniformly integrable case. By Lebesgue decomposition we reduce our consideration to the case when $Q \perp P$. Let us change the reference measure to $P' = \frac{P+Q}{2}$. The Radon-Nikodym derivative
$$X'_n = \left.\frac{dQ}{dP'}\right|_{\mathcal{F}_n} = \frac{2X_n}{1 + X_n}$$
is uniformly integrable with respect to P' and $X'_n \to X'$ a.e. P'. From the orthogonality $P \perp Q$ we know that there are disjoint sets E and E^c with $P(E) = 1$ and $Q(E^c) = 1$. Then
$$Q(A) = Q(A \cap E) + Q(A \cap E^c) = Q(A \cap E^c)$$
$$= 2P'(A \cap E^c) = \int_A 2 \mathbf{1}_E(\omega) dP'.$$

It is now seen that

$$X' = \frac{dQ}{dP'}\bigg|_{\mathcal{F}} = \begin{cases} 2 & \text{a.e. } Q \\ 0 & \text{a.e. } P \end{cases}$$

from which one concludes that

$$P\left\{\lim_{n\to\infty} X_n = 0\right\} = 1.$$

EXERCISE 5.4. In order to establish that a nonnegative martingale has an almost sure limit (which may not be an L_1 limit), show that we can assume, without loss of generality, that we are in the following situation:

$$\Omega = \otimes_{j=1}^{\infty} R, \quad \mathcal{F}_n = \sigma[x_1, x_2, \ldots, x_n], \quad X_j(\omega) = x_j.$$

The existence of a Q such that

$$\frac{dQ}{dP}\bigg|_{\mathcal{F}_n} = x_n$$

is essentially Kolmogorov's consistency theorem (Theorem 3.5). Now complete the proof.

REMARK 5.10. We shall give a more direct proof of almost sure convergence of an L_1-bounded martingale later on by means of the up-crossing inequality.

5.4. Stopping Times

The notion of stopping times that we studied in the context of Markov chains is important again in the context of martingales. In fact, the concept of stopping times is relevant whenever one has an ordered sequence of sub σ-fields and is concerned about conditioning with respect to them.

Let $(\Omega, \mathcal{F})$ be a measurable space and $\{\mathcal{F}_t : t \in T\}$ be a family of sub σ-fields. T is an ordered set, usually a set of real numbers or integers of the form $T = \{t : a \le t \le b\}$ or $\{t : t \ge a\}$. We will assume that $T = \{0, 1, 2, \ldots\}$, the set of nonnegative integers. The family $\mathcal{F}_n$ is assumed to be increasing with n. In other words,

$$\mathcal{F}_m \subset \mathcal{F}_n \quad \text{if } m < n.$$

An $\mathcal{F}$-measurable random variable $\tau(\omega)$ mapping $\Omega \to \{0, 1, \ldots, \infty\}$ is said to be a stopping time if for every $n \ge 0$ the set $\{\omega : \tau(\omega) \le n\} \in \mathcal{F}_n$. A stopping time may actually take the value ∞ on a nonempty subset of Ω.

The idea behind the definition of a stopping time, as we saw in the study of Markov chains, is that the decision to stop at time n can be based only on the information available up to that time.

EXERCISE 5.5. Show that the function $\tau(\omega) \equiv k$ is a stopping time for any admissible value of the constant k.

EXERCISE 5.6. Show that if τ is a stopping time and $f : T \to T$ is a nondecreasing function that satisfies $f(t) \ge t$ for all $t \in T$, then $\tau'(\omega) = f(\tau(\omega))$ is again a stopping time.

EXERCISE 5.7. Show that if τ_1 and τ_2 are stopping times, so are $\max(\tau_1, \tau_2)$ and $\min(\tau_1, \tau_2)$. In particular, any stopping time τ is an *increasing* limit of bounded stopping times $\tau_n(\omega) = \min(\tau(\omega), n)$.

To every stopping time τ we associate a stopped σ-field $\mathcal{F}_\tau \subset \mathcal{F}$ defined by

(5.5) $\qquad \mathcal{F}_\tau = \{A : A \in \mathcal{F} \text{ and } A \cap \{\omega : \tau(\omega) \le n\} \in \mathcal{F}_n \text{ for every } n\}.$

This should be thought of as the information available up to the stopping time τ. In other words, events in $\mathcal{F}_\tau$ correspond to questions that can be answered with a yes or no if we stop observing the process at time τ.

EXERCISE 5.8. Verify that for any stopping time τ, $\mathcal{F}_\tau$ is indeed a sub σ-field, i.e., is closed under countable unions and complementations. If $\tau(\omega) \equiv k$, then $\mathcal{F}_\tau \equiv \mathcal{F}_k$. If $\tau_1 \le \tau_2$ are stopping times, then $\mathcal{F}_{\tau_1} \subset \mathcal{F}_{\tau_2}$. Finally, if τ is a stopping time, then it is $\mathcal{F}_\tau$-measurable.

EXERCISE 5.9. If $X_n(\omega)$ is a sequence of measurable functions on $(\Omega, \mathcal{F})$ such that for every $n \in T$, X_n is $\mathcal{F}_n$-measurable, then on the set $\{\omega : \tau(\omega) < \infty\}$, which is an $\mathcal{F}_\tau$-measurable set, the function $X_\tau(\omega) = X_{\tau(\omega)}(\omega)$ is $\mathcal{F}_\tau$-measurable.

The following theorem, called *Doob's optional stopping theorem*, is one of the central facts in the theory of martingale sequences:

THEOREM 5.11 (Optional Stopping Theorem) *Let $\{X_n : n \ge 0\}$ be a sequence of random variables defined on a probability space $(\Omega, \mathcal{F}, P)$, which is a martingale sequence with respect to the filtration $(\Omega, \mathcal{F}_n, P)$, and let $0 \le \tau_1 \le \tau_2 \le C$ be two bounded stopping times. Then*

$$E[X_{\tau_2} \mid \mathcal{F}_{\tau_1}] = X_{\tau_1} \quad \text{a.e.}$$

PROOF: Since $\mathcal{F}_{\tau_1} \subset \mathcal{F}_{\tau_2} \subset \mathcal{F}_C$, it is sufficient to show that for any martingale $\{X_n\}$

(5.6) $\qquad E[X_k \mid \mathcal{F}_\tau] = X_\tau \quad \text{a.e.},$

provided τ is a stopping time bounded by the integer k. To see this we note that in view of Exercise 4.10,

$$E[X_k \mid \mathcal{F}_{\tau_1}] = E[E[X_k \mid \mathcal{F}_{\tau_2}] \mid \mathcal{F}_{\tau_1}],$$

and if (5.6) holds, then

$$E[X_{\tau_2} \mid \mathcal{F}_{\tau_1}] = X_{\tau_1} \quad \text{a.e.}$$

Let $A \in \mathcal{F}_\tau$. If we define $E_j = \{\omega : \tau(\omega) = j\}$, then $\Omega = \bigcup_1^k E_j$ is a disjoint union. Moreover, $A \cap E_j \in \mathcal{F}_j$ for every $j = 1, 2, \ldots, k$. By the martingale property

$$\int_{A \cap E_j} X_k \, dP = \int_{A \cap E_j} X_j \, dP = \int_{A \cap E_j} X_\tau \, dP$$

and summing over $j = 1, 2, \ldots, k$ gives

$$\int_A X_k \, dP = \int_A X_\tau \, dP$$

5.4. STOPPING TIMES

for every $A \in \mathcal{F}_\tau$, and we are done. □

REMARK 5.11. In particular, if X_n is a martingale sequence and τ is a bounded stopping time, then $E[X_\tau] = E[X_0]$. This property, obvious for constant times, has now been extended to bounded stopping times. In a "fair" game, a policy to quit at an "opportune" time gives no advantage to the gambler so long as he or she cannot foresee the future.

EXERCISE 5.10. The property extends to sub- or supermartingales. For example, if X_n is a submartingale, then for any two bounded stopping times $\tau_1 \leq \tau_2$, we have
$$E[X_{\tau_2} \mid \mathcal{F}_{\tau_1}] \geq X_{\tau_1} \quad \text{a.e.}$$
One cannot use the earlier proof directly, but one can reduce it to the martingale case by applying the Doob decomposition theorem.

EXERCISE 5.11. Boundedness is important. Take $X_0 = 0$ and
$$X_n = \xi_1 + \xi_2 + \cdots + \xi_n \quad \text{for } n \geq 1$$
where ξ_i are independent identically distributed random variables taking the values ± 1 with probability $\frac{1}{2}$. Let $\tau = \inf\{n : X_n = 1\}$. Then τ is a stopping time, $P[\tau < \infty] = 1$, but τ is not bounded. $X_\tau = 1$ with probability 1 and trivially $E[X_\tau] = 1 \neq 0$.

EXERCISE 5.12. It does not mean that we can never consider stopping times that are unbounded. Let τ be an unbounded stopping time. For every k, $\tau_k = \min(\tau, k)$ is a bounded stopping time and $E[X_{\tau_k}] = 0$ for every k. As $k \to \infty$, $\tau_k \uparrow \tau$ and $X_{\tau_k} \to X_\tau$. If we can establish uniform integrability of X_{τ_k}, we can pass to the limit. In particular, if $S(\omega) = \sup_{0 \leq n \leq \tau(\omega)} |X_n(\omega)|$ is integrable, then $\sup_k |X_{\tau_k}(\omega)| \leq S(\omega)$ and therefore $E[X_\tau] = 0$.

EXERCISE 5.13. Use a similar argument to show that if
$$S(\omega) = \sup_{0 \leq k \leq \tau_2(\omega)} |X_k(\omega)|$$
is integrable, then for any $\tau_1 \leq \tau_2$
$$E[X_{\tau_2} \mid \mathcal{F}_{\tau_1}] = X_{\tau_1} \quad \text{a.e.}$$

EXERCISE 5.14. The previous exercise needs the fact that if $\tau_n \uparrow \tau$ are stopping times, then
$$\sigma\left\{\bigcup_n \mathcal{F}_{\tau_n}\right\} = \mathcal{F}_\tau .$$
Prove it.

EXERCISE 5.15. Let us go back to the earlier exercise (Exercise 5.11) where we had
$$X_n = \xi_1 + \xi_2 + \cdots + \xi_n$$
as a sum of n independent random variables taking the values ± 1 with probability $\frac{1}{2}$. Show that if τ is a stopping time with $E[\tau] < \infty$, then $S(\omega) = \sup_{1 \leq n \leq \tau(\omega)} |X_n(\omega)|$ is square-integrable and therefore $E[X_\tau] = 0$.

Hint. Use the fact that $X_n^2 - n$ is a martingale.

5.5. Up-crossing Inequality

The following inequality due to Doob, which controls the oscillations of a martingale sequence, is very useful for proving the almost sure convergence of L_1-bounded martingales directly. Let $\{X_j : 0 \leq j \leq n\}$ be a martingale sequence with $n+1$ terms. Let us take two real numbers $a < b$. An up-crossing is a pair of terms X_k and X_l, with $k < l$, for which $X_k \leq a < b \leq X_l$. Starting from X_0, we locate the first term that is at most a and then the first term following it that is at least b. This is the first up-crossing. In our martingale sequence there will be a certain number of completed up-crossings (of course over disjoint intervals) and then at the end we may be in the middle of an up-crossing or may not even have started on one because we are still on the way down from a level above b to one below a. In any case, there will be a certain number $U(a, b)$ of completed up-crossings. Doob's up-crossing inequality gives a uniform upper bound on the expected value of $U(a, b)$ in terms of $E[|X_n|]$, i.e., one that does not depend otherwise on n.

THEOREM 5.12 (Doob's Up-crossing Inequality) *For any n,*

$$(5.7) \quad E[U(a,b)] \leq \frac{1}{b-a} E[a - X_n]^+ \leq \frac{1}{b-a} [|a| + E[|X_n|]] .$$

PROOF: Let us define recursively

$$\tau_1 = n \wedge \inf\{k : X_k \leq a\}$$
$$\tau_2 = n \wedge \inf\{k : k \geq \tau_1, X_k \geq b\}$$
$$\vdots$$
$$\tau_{2k} = n \wedge \inf\{k : k \geq \tau_{2k-1}, X_k \geq b\}$$
$$\tau_{2k+1} = n \wedge \inf\{k : k \geq \tau_{2k}, X_k \leq a\} .$$

Since $\tau_k \geq \tau_{k-1} + 1$, $\tau_n = n$. Consider the quantity

$$D(\omega) = \sum_{j=1}^{n} [X_{\tau_{2j}} - X_{\tau_{2j-1}}] ,$$

which could very well have lots of 0's at the end. In any case, the first few terms correspond to up-crossings, and each term is at least $(b-a)$ and there are $U(a, b)$ of them. Before the 0's begin there may be at most one nonzero term that is an incomplete up-crossing, i.e., when $\tau_{2\ell-1} < n = \tau_{2\ell}$ for some ℓ. It is then equal to $(X_n - X_{\tau_{2l-1}}) \geq X_n - a$ for some l. If, on the other hand, we end in the middle of a down-crossing, i.e., $\tau_{2\ell} < n = \tau_{2\ell+1}$, there is no incomplete up-crossing. Therefore,

$$D(\omega) \geq (b-a)U(a,b) + R_n(\omega)$$

with the remainder $R_n(\omega)$ satisfying

$$R_n(\omega) = 0 \qquad \text{if } \tau_{2\ell} < n = \tau_{2\ell+1}$$
$$\geq (X_n - a) \quad \text{if } \tau_{2\ell-1} < n = \tau_{2\ell} .$$

By the optional stopping theorem, $E[D(\omega)] = 0$. This gives the bound

$$E[U(a,b)] \leq \frac{1}{b-a} E[-R_n(\omega)] \leq \frac{1}{b-a} E[(a-X_n)^+]$$
$$\leq \frac{1}{b-a} E[|a-X_n|] \leq \frac{1}{b-a} E[|a| + |X_n|].$$

□

REMARK 5.12. In particular, if X_n is an L_1-bounded martingale, then the number of up-crossings of any interval $[a, b]$ is finite with probability 1. From Doob's inequality, the sequence X_n is almost surely bounded. It now follows by taking a countable number of intervals $[a, b]$ with rational endpoints that X_n has a limit almost surely. If X_n is uniformly integrable, then the convergence is in L_1 and then $X_n = E[X \mid \mathcal{F}_n]$. If we have a uniform L_p-bound on X_n, then $X \in L_p$ and $X_n \to X$ in L_p. All of our earlier results on the convergence of martingales now follow.

EXERCISE 5.16. For the proof it is sufficient that we have a supermartingale. In fact, we can change signs, and so a submartingale works just as well.

5.6. Martingale Transforms, Option Pricing

If X_n is a martingale with respect to $(\Omega, \mathcal{F}_n, P)$ and Y_n are the differences $X_n - X_{n-1}$, a martingale transform X'_n of X_n is given by the formula

$$X'_n = X'_{n-1} + a_{n-1} Y_n \quad \text{for } n \geq 1$$

where a_{n-1} is $\mathcal{F}_{n-1}$-measurable and has enough integrability assumptions to make $a_{n-1} Y_n$ integrable. An elementary calculation shows that

$$E[X'_n \mid \mathcal{F}_{n-1}] = X'_{n-1},$$

making X'_n a martingale as well. X'_n is called a *martingale transform* of X_n. The interpretation is if we have a fair game, we can choose the size and side of our bet at each stage based on the prior history and the game will continue to be fair. It is important to note that X_n may be sums of independent random variables with mean 0. But the independence of the increments may be destroyed, and X'_n will in general no longer have the independent increments property.

EXERCISE 5.17. Suppose $X_n = \xi_1 + \xi_2 + \cdots + \xi_n$, where ξ_j are independent random variables taking the values ± 1 with probability $\frac{1}{2}$. Let X'_n be the martingale transform given by

$$X'_n = \sum_{j=1}^{n} a_{j-1}(\omega) \xi_j$$

where a_j is $\mathcal{F}_j$-measurable, $\mathcal{F}_j$ being the σ-field generated by $\xi_1, \xi_2, \ldots, \xi_j$. Calculate $E\{[X'_n]^2\}$.

Suppose X_n is a sequence of nonnegative random variables that represent the value of a security that is traded in the market place at a price that is X_n for day n and changes overnight between day n and day $n + 1$ from X_n to X_{n+1}. We could at the end of day n, based on any information $\mathcal{F}_n$ that is available to us at the end of that day, be either long or short on the security. The quantity $a_n(\omega)$ is the number of shares that we choose to own overnight between day n and day $n+1$ and that could be a function of all the information available to us up to that point. Positive values of a_n represent long positions and negative values represent short positions. Our gain or loss overnight is given by $a_n(X_{n+1} - X_n)$, and the cumulative gain (loss) is the transform

$$X'_n - X'_0 = \sum_{j=1}^{n} a_{j-1}(X_j - X_{j-1}).$$

A contingent claim (European option) is really a gamble or a bet based on the value of X_N at some terminal date N. The nature of the claim is that there is a function $f(x)$ such that if the security trades on that day at a price x, then the claim pays an amount of $f(x)$. A *call* is an option to buy at a certain price a and the payoff is $f(x) = (x-a)^+$, whereas a *put* is an option to sell at a fixed price a and therefore has a payoff function $f(x) = (a-x)^+$.

Replicating a claim, if it is possible at all, is determining $a_0, a_1, \ldots, a_{N-1}$ and the initial value V_0 such that the transform

$$V_N = V_0 + \sum_{j=0}^{N-1} a_j(X_{j+1} - X_j)$$

at time N equals the claim $f(X_N)$ under every conceivable behavior of the price movements $X_1, X_2, \ldots, X_N$. If the claim can be exactly replicated starting from an initial capital of V_0, then V_0 becomes the price of that option. Anyone could sell the option at that price, use the proceeds as capital, and follow the strategy dictated by the coefficients $a_0, a_1, \ldots, a_{N-1}$ and have *exactly* enough to pay off the claim at time N. Here we are ignoring transaction costs as well as interest rates. It is not always true that a claim can be replicated.

Let us assume for simplicity that the stock prices are always some nonnegative integral multiples of some unit. The set of possible prices can then be taken to be the set of nonnegative integers. Let us make a crucial assumption that if the price on some day is x, the price on the next day is $x \pm 1$. It has to move up or down a notch. It cannot jump two or more steps or even stay the same. When the stock price hits 0 we assume that the company goes bankrupt and the stock stays at 0 forever. In all other cases, from day to day, it always moves either up or down a notch.

Let us value the claim f for one period. If the price at day $N-1$ is $x \neq 0$ and we have assets c on hand and invest in a shares, we will end up on day N with either assets of $c + a$ and a claim of $f(x+1)$ or assets of $c - a$ with a claim of $f(x-1)$. In order to make sure that we break even in either case, we need

$$f(x+1) = c + a, \quad f(x-1) = c - a,$$

and, solving for a and c, we get
$$c(x) = \frac{1}{2}[f(x+1) + f(x-1)], \qquad a(x) = \frac{1}{2}[f(x+1) - f(x-1)].$$

The value of the claim with one day left is
$$V_{N-1}(x) = \begin{cases} \frac{1}{2}[f(x-1) + f(x+1)] & \text{if } x \geq 1 \\ f(0) & \text{if } x = 0, \end{cases}$$
and we can proceed by iteration
$$V_{j-1}(x) = \begin{cases} \frac{1}{2}[V_j(x-1) + V_j(x+1)] & \text{if } x \geq 1 \\ V_j(0) & \text{if } x = 0 \end{cases}$$
for $j \geq 1$ till we arrive at the value $V_0(x)$ of the claim at time 0 and price x. The corresponding values of $a = a_{j-1}(x) = \frac{1}{2}[V_j(x+1) - V_j(x-1)]$ give us the number of shares to hold between day $j-1$ and j if the current price at time $j-1$ equals x.

REMARK 5.13. The important fact is that the value is determined by arbitrage and is unaffected by the actual movement of the price so long as it is compatible with the model.

REMARK 5.14. The value does not depend on any statistical assumptions on the various probabilities of transitions of price levels between successive days.

REMARK 5.15. However, the value can be interpreted as the expected value
$$V_0(x) = E^{P_x}\{f(X_N)\}$$
where P_x is the random walk starting at x with probability $\frac{1}{2}$ for transitions up or down a level, which is absorbed at 0.

REMARK 5.16. P_x can be characterized as the unique probability distribution of $(X_0, X_1, \ldots, X_N)$ such that $P_x[X_0 = x] = 1$, $P_x[|X_j - X_{j-1}| = 1 \mid X_{j-1} \geq 1] = 1$ for $1 \leq j \leq N$ and X_j is a martingale with respect to $(\Omega, \mathcal{F}_j, P_x)$ where $\mathcal{F}_j$ is generated by $X_0, X_1, \ldots, X_j$.

EXERCISE 5.18. It is not necessary for the argument that the set of possible price levels be equally spaced. If we make the assumption that, for each price level $x > 0$, the price on the following day can take only one of two possible values $h(x) > x$ and $l(x) < x$ with a possible bankruptcy if the level 0 is reached, a similar analysis can be worked out. Carry it out.

5.7. Martingales and Markov Chains

One of the ways of specifying the joint distribution of a sequence $X_0, X_1, \ldots, X_n$ of random variables is to specify the distribution of X_0 and, for each $j \geq 1$, specify the conditional distribution of X_j given the σ-field $\mathcal{F}_{j-1}$ generated by $X_0, X_1, \ldots, X_{j-1}$. Equivalently, instead of the conditional distributions one can specify the conditional expectations $E[f(X_j) \mid \mathcal{F}_{j-1}]$ for $1 \leq j \leq n$. Let us write
$$h_{j-1}(X_0, X_1, \ldots, X_{j-1}) = E[f(X_j) \mid \mathcal{F}_{j-1}] - f(X_{j-1})$$

so that, for $1 \leq j \leq n$,
$$E\big[\{f(X_j) - f(X_{j-1}) - h_{j-1}(X_0, X_1, \ldots, X_{j-1})\} \mid \mathcal{F}_{j-1}\big] = 0$$
or
$$Z_j^f = f(X_j) - f(X_0) - \sum_{i=1}^{j} h_{i-1}(X_0, X_1, \ldots, X_{i-1})$$
is a martingale for every f. It is not difficult to see that the specification of $\{h_i\}$ for each f is enough to determine all the successive conditional expectations and therefore the conditional distributions. If, in addition, the initial distribution of X_0 is specified, then the distribution of $(X_0, X_1, \ldots, X_n)$ is completely determined.

If, for each j and f, the corresponding $h_{j-1}(X_0, X_1, \ldots, X_{j-1})$ is a function $h_{j-1}(X_{j-1})$ of X_{j-1} only, then the distribution of $(X_0, X_1, \ldots, X_n)$ is Markov and the transition probabilities are seen to be given by the relation
$$h_{j-1}(X_{j-1}) = E\big[[f(X_j) - f(X_{j-1})] \mid \mathcal{F}_{j-1}\big]$$
$$= \int [f(y) - f(X_{j-1})]\pi_{j-1,j}(X_{j-1}, dy).$$

In the case of a stationary Markov chain, the relationship is
$$h_{j-1}(X_{j-1}) = h(X_{j-1}) = E\big[[f(X_j) - f(X_{j-1})] \mid \mathcal{F}_{j-1}\big]$$
$$= \int [f(y) - f(X_{j-1})]\pi(X_{j-1}, dy).$$

If we introduce the linear transformation (transition operator)

(5.8) $$(\Pi f)(x) = \int f(y)\pi(x, dy),$$

then
$$h(x) = ([\Pi - I]f)(x).$$

REMARK 5.17. In the case of a Markov chain on a countable state space,
$$(\Pi f)(x) = \sum_y \pi(x, y) f(y)$$
and
$$h(x) = [\Pi - I]f(x) = \sum_y [f(y) - f(x)]\pi(x, y).$$

REMARK 5.18. The measure P_x on the space $(\Omega, \mathcal{F})$ of sequences $\{x_j : j \geq 0\}$ from the state space X that corresponds to the Markov process with transition probability $\pi(x, dy)$ and initial state x can be characterized as the unique measure on $(\Omega, \mathcal{F})$ such that
$$P_x\{\omega : x_0 = x\} = 1,$$
and for every bounded measurable function f defined on the state space X,
$$f(x_n) - f(x_0) - \sum_{j=1}^{n} h(x_{j-1})$$

is a martingale with respect to $(\Omega, \mathcal{F}_n, P_x)$, where

$$h(x) = \int_X [f(y) - f(x)]\pi(x, dy).$$

Let $A \subset X$ be a measurable subset and let $\tau_A = \inf\{n \geq 0 : x_n \in A\}$ be the first entrance time into A. It is easy to see that τ_A is a stopping time. It need not always be true that $P_x\{\tau_A < \infty\} = 1$. But $U_A(x) = P_x\{\tau_A < \infty\}$ is a well-defined measurable function of x that satisfies $0 \leq U(x) \leq 1$ for all x and is the exit probability from the set A^c. By its very definition $U_A(x) \equiv 1$ on A, and if $x \notin A$, by the Markov property,

$$U_A(x) = \pi(x, A) + \int_{A^c} U_A(y)\pi(x, dy) = \int_X U_A(y)\pi(x, dy).$$

In other words, U_A satisfies $0 \leq U_A \leq 1$ and is a solution of

(5.9) $$\begin{aligned} (\Pi - I)V &= 0 \quad \text{on } A^c \\ V &= 1 \quad \text{on } A. \end{aligned}$$

THEOREM 5.13 *Among all nonnegative solutions V of equation (5.9) $U_A(x) = P_x\{\tau_A < \infty\}$ is the smallest. If $U_A(x) = 1$, then any bounded solution of the equation*

(5.10) $$\begin{aligned} (\Pi - I)V &= 0 \quad \text{on } A^c \\ V &= f \quad \text{on } A \end{aligned}$$

is equal to

(5.11) $$V(x) = E^{P_x}\{f(x_{\tau_A})\}.$$

In particular, if $U_A(x) = 1$ for all $x \notin A$, then any bounded solution V of equation (5.10) is unique and is given by formula (5.11).

PROOF: First we establish that any nonnegative solution V of (5.9) dominates U_A. Let us replace V by $W = \min(V, 1)$. Then $0 \leq W \leq 1$ everywhere, $W(x) = 1$ for $x \in A$, and for $x \notin A$,

$$(\Pi W)(x) = \int_X W(y)\pi(x, dy) \leq \int_X V(y)\pi(x, dy) = V(x).$$

Since $\Pi W \leq 1$ as well, we conclude that $\Pi W \leq W$ on A^c. On the other hand, it is obvious that $\Pi W \leq 1 = W$ on A. Since we have shown that $\Pi W \leq W$ everywhere, it follows that $\{W(x_n)\}$ is a supermartingale with respect to $(\Omega, \mathcal{F}_n, P_x)$. In particular, for any bounded stopping time τ,

$$E^{P_x}\{W(x_\tau)\} \leq E^{P_x}\{W(x_0)\} = W(x).$$

While we cannot take $\tau = \tau_A$ (since τ_A may not be bounded), we can always take $\tau = \tau_N = \min(\tau_A, N)$ to conclude

$$E^{P_x}\{W(x_{\tau_N})\} \leq E^{P_x}\{W(x_0)\} = W(x).$$

Let $N \to \infty$. On the set $\{\omega : \tau_A(\omega) < \infty\}$, $\tau_N \uparrow \tau_A$ and $W(x_{\tau_N}) \to W(x_{\tau_A}) = 1$. Since W is nonnegative and bounded,

$$W(x) \geq \limsup_{N \to \infty} E^{P_x}\{W(x_{\tau_N})\} \geq \limsup_{N \to \infty} \int_{\tau_A < \infty} W(x_{\tau_N}) dP_x$$
$$= P_x\{\tau_A < \infty\} = U_A(x).$$

Since $V(x) \geq W(x)$, it follows that $V(x) \geq U_A(x)$.

For a bounded solution V of (5.10), let us define $h = (\Pi - I)V$, which will be a function vanishing on A^c. We know that

$$V(x_n) - V(x_0) - \sum_{j=1}^{n} h(x_{j-1})$$

is a martingale with respect to $(\Omega, \mathcal{F}_n, P_x)$, and let us use the stopping theorem with $\tau_N = \min(\tau_A, N)$. Since $h(x_{j-1}) = 0$ for $j \leq \tau_A$, we obtain

$$V(x) = E^{P_x}\{V(x_{\tau_N})\}.$$

If we now make the assumption that $U_A(x) = P_x\{\tau_A < \infty\} = 1$, let $N \to \infty$, and use the bounded convergence theorem, it is easy to see that

$$V(x) = E^{P_x}\{f(x_{\tau_A})\},$$

which proves (5.11) and the rest of the theorem. $\square$

Such arguments are powerful tools for studying the qualitative properties of Markov chains. Solutions to equations of the type $[\Pi - I]V = f$ are often easily constructed. They can be used to produce martingales, submartingales, or supermartingales that have certain behavior, and that, in turn, implies certain qualitative behavior of the Markov chain. We will now see several illustrations of this method.

EXAMPLE 5.1. Consider the symmetric simple random walk in one dimension.

We know from recurrence that the random walk exits the interval $(-R, R)$ in a finite time. But we want to get some estimates on the exit time τ_R. Consider the function $u(x) = \cos \lambda x$. The function $f(x) = [\Pi u](x)$ can be calculated and

$$f(x) = \frac{1}{2}[\cos \lambda(x - 1) + \cos \lambda(x + 1)] = \cos \lambda \cos \lambda x = (\cos \lambda) u(x).$$

If $\lambda < \frac{\pi}{2R}$, then $\cos \lambda x \geq \cos \lambda R > 0$ in $[-R, R]$. Consider $Z_n = e^{\sigma n} \cos \lambda x_n$ with $\sigma = -\log \cos \lambda$;

$$E^{P_x}\{Z_n \mid \mathcal{F}_{n-1}\} = e^{\sigma n} f(x_{n-1}) = e^{\sigma n} \cos \lambda \cos \lambda x_{n-1} = Z_{n-1}.$$

If τ_R is the exit time from the interval $(-R, R)$, for any N we have

$$E^{P_x}\{Z_{\tau_R \wedge N}\} = E^{P_x}\{Z_0\} = \cos \lambda x.$$

Since $\sigma > 0$ and $\cos \lambda x \geq \cos \lambda R > 0$ for $x \in [-R, R]$, if R is an integer, we can claim that

$$E^{P_x}\{e^{\sigma[\tau_R \wedge N]}\} \leq \frac{\cos \lambda x}{\cos \lambda R}.$$

5.7. MARTINGALES AND MARKOV CHAINS

Since the estimate is uniform, we can let $N \to \infty$ to get the estimate

$$E^{P_x}\{e^{\sigma \tau_R}\} \leq \frac{\cos \lambda x}{\cos \lambda R}.$$

EXERCISE 5.19. Can you prove equality above? What is the range of validity of the equality? Is $E^{P_x}\{e^{\sigma \tau_R}\} < \infty$ for all $\sigma > 0$?

EXAMPLE 5.2. Let us make life slightly more complicated by taking a Markov chain in $\mathbb{Z}^d$ with transition probabilities

$$\pi(x, y) = \begin{cases} \frac{1}{2d} + \delta(x, y) & \text{if } |x - y| = 1 \\ 0 & \text{if } |x - y| \neq 0, \end{cases}$$

so that we have slightly perturbed the random walk with perhaps even a possible bias.

Exact calculations like those in Example 5.1 are of course no longer possible. Let us try to estimate again the exit time from a ball of radius R. For $\sigma > 0$, consider the function

$$F(x) = \exp\left[\sigma \sum_{i=1}^{d} |x_i|\right]$$

defined on $\mathbb{Z}^d$. We can get an estimate of the form

$$(\Pi F)(x_1, x_2, \ldots, x_d) \geq \theta F(x_1, x_2, \ldots, x_d)$$

for some choices of $\sigma > 0$ and $\theta > 1$ that may depend on R. Now proceed as in Example 5.1.

EXAMPLE 5.3. We can use these methods to show that the random walk is transient in dimension $d \geq 3$.

For $0 < \alpha < d - 2$ consider the function $V(x) = \frac{1}{|x|^\alpha}$ for $x \neq 0$ with $V(0) = 1$. An approximate calculation of $(\Pi V)(x)$ yields, for sufficiently large $|x|$ (i.e., $|x| \geq L$ for some L), the estimate

$$(\Pi V)(x) - V(x) \leq 0.$$

If we start from an x with $|x| > L$ and take τ_L to be the first entrance time into the ball of radius L, we get by the stopping theorem the inequality

$$E^{P_x}\{V(x_{\tau_L \wedge N})\} \leq V(x).$$

If $\tau_L \leq N$, then $|x_{\tau_L}| \leq L$. In any case, $V(x_{\tau_L \wedge N}) \geq 0$. Therefore,

$$P_x\{\tau_L \leq N\} \leq \frac{V(x)}{\inf_{|y| \leq L} V(y)}$$

is valid uniformly in N. Letting $N \to \infty$

$$P_x\{\tau_L < \infty\} \leq \frac{V(x)}{\inf_{|y| \leq L} V(y)}.$$

If we let $|x| \to \infty$, keeping L fixed, we see the transience. Note that recurrence implies that $P_x\{\tau_L < \infty\} = 1$ for all x. The proof of transience really only required a function V defined for large $|x|$ that was strictly positive for each x, went to 0 as $|x| \to \infty$, and had the property $(\Pi V)(x) \leq V(x)$ for large values of $|x|$.

EXAMPLE 5.4. We will now show that the random walk is recurrent in $d = 2$.

This is harder because the recurrence of random walk in $d = 2$ is right on the border. We want to construct a function $V(x) \to \infty$ as $|x| \to \infty$ that satisfies $(\Pi V)(x) \leq V(x)$ for large $|x|$. If we succeed, then we can estimate by a stopping argument the probability that the chain starting from a point x in the annulus $\ell < |x| < L$ exits at the outer circle before getting inside the inner circle:

$$P_x\{\tau_L < \tau_\ell\} \leq \frac{V(x)}{\inf_{|y| \geq L} V(y)}.$$

We also have, for every L,

$$P_x\{\tau_L < \infty\} = 1.$$

This proves that $P_x\{\tau_\ell < \infty\} = 1$, thereby proving recurrence. The natural candidate is $F(x) = \log|x|$ for $x \neq 0$. A computation yields

$$(\Pi F)(x) - F(x) \leq \frac{C}{|x|^4},$$

which does not quite make it. On the other hand, if $U(x) = |x|^{-1}$, for large values of $|x|$,

$$(\Pi U)(x) - U(x) \geq \frac{c}{|x|^3}$$

for some $c > 0$. The choice of $V(x) = F(x) - CU(x) = \log x - \frac{C}{|x|}$ works with any $C > 0$.

EXAMPLE 5.5. We can use these methods for proving positive recurrence as well.

Suppose X is a countable set and we can find $V \geq 0$, a finite set F, and a constant $C \geq 0$ such that

$$(\Pi V)(x) - V(x) \leq \begin{cases} -1 & \text{for } x \notin F \\ C & \text{for } x \in F. \end{cases}$$

Let $U = \Pi V - V$, and we have

$$-V(x) \leq E^{P_x}\{V(x_n) - V(x)\} = E^{P_x}\left\{\sum_{j=1}^n U(x_{j-1})\right\}$$

$$\leq E^{P_x}\left\{\sum_{j=1}^n C\mathbf{1}_F(x_{j-1}) - \sum_{j=1}^n \mathbf{1}_{F^c}(x_{j-1})\right\}$$

$$= -E^{P_x}\left\{\sum_{j=1}^n [1 - (1+C)\mathbf{1}_F(x_{j-1})]\right\}$$

$$= -n + (1+C) \sum_{j=1}^{n} \sum_{y \in F} \pi^n(x, y)$$

$$= -n + o(n) \quad \text{as } n \to \infty$$

if the process is not positive recurrent. This is a contradiction.

For instance, if $X = \mathbb{Z}$, the integers, and we have a little bit of bias towards the origin in the random walk

$$\pi(x, x+1) - \pi(x, x-1) \geq \frac{a}{|x|} \quad \text{if } x \leq -\ell$$

$$\pi(x, x-1) - \pi(x, x+1) \geq \frac{a}{|x|} \quad \text{if } x \geq \ell$$

with $V(x) = x^2$, for $x \geq \ell$,

$$(\Pi V)(x) \leq (x+1)^2 \frac{1}{2}\left(1 - \frac{a}{|x|}\right) + (x-1)^2 \frac{1}{2}\left(1 + \frac{a}{|x|}\right) = x^2 + 1 - 2a.$$

If $a > \frac{1}{2}$, we can multiply V by a constant and it works.

EXERCISE 5.20. What happens when

$$\pi(x, x+1) - \pi(x, x-1) = -\frac{1}{2x}$$

for $|x| \geq 10$? (See Exercise 4.17.)

EXAMPLE 5.6. Let us return to our example of a branching process in Example 4.4. We see from the relation

$$E[X_{n+1} \mid \mathcal{F}_n] = m X_n$$

that $\frac{X_n}{m^n}$ is a martingale. If $m < 1$, we saw before quite easily that the population becomes extinct. If $m = 1$, X_n is a martingale. Since it is nonnegative, it is L_1 bounded and must have an almost sure limit as $n \to \infty$. Since the population is an integer, this means that the size eventually stabilizes. The limit can only be 0 because the population cannot stabilize at any other size. If $m > 1$, there is a probability $0 < q < 1$ such that $P[X_n \to 0 \mid X_0 = 1] = q$. We can show that with probability $1 - q$, $X_n \to \infty$. To see this, consider the function $u(x) = q^x$. In the notation of Example 4.4

$$E[q^{X_{n+1}} \mid \mathcal{F}_n] = \left[\sum q^j p_j\right]^{X_n} = [P(q)]^{X_n} = q^{X_n}$$

so that q^{X_n} is a nonnegative martingale. It then has an almost sure limit, which can only be 0 or 1. If q is the probability that it is 1, i.e., that $X_n \to 0$, then $1 - q$ is the probability that it is 0, i.e., that $X_n \to \infty$.

CHAPTER 6

Stationary Stochastic Processes

6.1. Ergodic Theorems

A stationary stochastic process is a collection $\{\xi_n : n \in \mathbb{Z}\}$ of random variables with values in some space $(X, \mathcal{B})$ such that the joint distribution of $(\xi_{n_1}, \xi_{n_2}, \ldots, \xi_{n_k})$ is the same as that of $(\xi_{n_1+n}, \xi_{n_2+n}, \ldots, \xi_{n_k+n})$ for every choice of $k \geq 1$ and $n, n_1, \ldots, n_k \in \mathbb{Z}$. Assuming that the space $(X, \mathcal{B})$ is reasonable and Kolmogorov's consistency theorem applies, we can build a measure P on the countable product space Ω of sequences $\{x_n : n \in \mathbb{Z}\}$ with values in X, defined for sets in the product σ-field $\mathcal{F}$. On the space Ω there is the natural shift defined by $(T\omega)(n) = x_{n+1}$ for ω with $\omega(n) = x_n$. The random variables $x_n(\omega) = \omega(n)$ are essentially equivalent to $\{\xi_n\}$. The stationarity of the process is reflected in the invariance of P with respect to T, i.e., $PT^{-1} = P$. We can without being specific consider a space Ω, a σ-field $\mathcal{F}$, a one-to-one invertible measurable map $T : \Omega \to \Omega$ with a measurable inverse T^{-1}, and, finally, a probability measure P on $(\Omega, \mathcal{F})$ that is T-invariant, i.e., $P(T^{-1}A) = P(A)$ for every $A \in \mathcal{F}$. One says that P is an *invariant measure* for T or T is a *measure-preserving transformation* for P. If we have a measurable map from $\xi : (\Omega, \mathcal{F}) \to (X, \mathcal{B})$, then it is easily seen that $\xi_n(\omega) = \xi(T^n\omega)$ defines a stationary stochastic process. The study of stationary stochastic processes is then more or less the same as the study of measure-preserving (i.e., probability-preserving) transformations.

The basic transformation $T : \Omega \to \Omega$ induces a linear transformation U on the space of functions defined on Ω by the rule $(Uf)(\omega) = f(T\omega)$. Because T is measure preserving, it is easy to see that

$$\int_\Omega f(\omega)dP = \int_\Omega f(T\omega)dP = \int_\Omega (Uf)(\omega)dP$$

as well as

$$\int_\Omega |f(\omega)|^p dP = \int_\Omega |f(T\omega)|^p dP = \int_\Omega |(Uf)(\omega)|^p dP.$$

In other words, U acts as an isometry (i.e., norm-preserving linear transformation) on the various L_p-spaces for $1 \leq p < \infty$, and, in fact, it is an isometry on L_∞ as well. Moreover, the transformation induced by T^{-1} is the inverse of U so that U is also invertible; in particular, U is unitary (or orthogonal) on L_2. This means it preserves the inner product $\langle \cdot, \cdot \rangle$,

$$\langle f, g \rangle = \int f(\omega) g(\omega) dP = \int f(T\omega) g(T\omega) dP = \langle Uf, Ug \rangle.$$

Of course, our linear transformation U is very special and satisfies $U1 = 1$ and $U(fg) = (Uf)(Ug)$.

A basic theorem known as the *ergodic theorem* asserts that

THEOREM 6.1 *For any $f \in L_1(P)$ the limit*

$$\lim_{n \to \infty} \frac{f(\omega) + f(T\omega) + \cdots + f(T^{n-1}\omega)}{n} = g(\omega)$$

exists for almost all ω with respect to P as well as in $L_1(P)$. Moreover, if $f \in L_p$ for some p satisfying $1 < p < \infty$, then the function $g \in L_p$ and the convergence takes place in that L_p. Moreover, the limit $g(\omega)$ is given by the conditional expectation

$$g(\omega) = E^P[f \mid \mathcal{I}]$$

where the σ-field $\mathcal{I}$, called the invariant σ-field, *is defined as*

$$\mathcal{I} = \{A : TA = A\}.$$

PROOF: First we prove the convergence in the various L_p-spaces. These are called *mean ergodic theorems*. The easiest situation to prove is when $p = 2$. Let us define

$$H_0 = \{f : f \in H, Uf = f\} = \{f : f \in H, f(T\omega) = f(\omega)\}.$$

Since H_0 contains constants, it is a closed nontrivial subspace of $H = L_2(P)$ of dimension at least 1. Since U is unitary, $Uf = f$ if and only if $U^{-1}f = U^*f = f$ where U^* is the adjoint of U. The orthogonal complement $H_0^\perp$ can be defined as

$$H_0^\perp = \{g : \langle g, f \rangle = 0 \, \forall f : U^*f = f\} = \overline{\text{Range}(I - U)H}.$$

Clearly, if we let

$$A_n f = \frac{f + Uf + \cdots + U^{n-1}f}{n},$$

then $\|A_n f\|_2 \leq \|f\|_2$ for every $f \in H$ and $A_n f = f$ for every n and $f \in H_0$. Therefore, for $f \in H_0$, $A_n f \to f$ as $n \to \infty$. On the other hand, if $f = (I - U)g$, $A_n f = \frac{g - U^n g}{n}$ and $\|A_n f\|_2 \leq \frac{2\|g\|_2}{n} \to 0$ as $n \to \infty$. Since $\|A_n\| \leq 1$, it follows that $A_n f \to 0$ as $n \to \infty$ for every $f \in H_0^\perp = \overline{\text{Range}(I - U)H}$; see Exercise 6.1. If we denote by π the orthogonal projection from $H \to H_0$, we see that $A_n f \to \pi f$ as $n \to \infty$ for every $f \in H$, establishing the L_2-ergodic theorem.

There is an alternate characterization of H_0. Functions f in H_0 are invariant under T, i.e., have the property that $f(T\omega) = f(\omega)$. For any invariant function f the level sets $\{\omega : a < f(\omega) < b\}$ are invariant under T. We can, therefore, talk about invariant sets $\{A : A \in \mathcal{F}, T^{-1}A = A\}$. Technically, we should allow ourselves to differ by sets of measure 0, and one defines $\mathcal{I} = \{A : P(A \Delta T^{-1}A) = 0\}$ as the σ-field of almost invariant sets.

Nothing is therefore lost by taking $\mathcal{I}$ to be the σ-field of invariant sets. We can identify the orthogonal projection π as (see Exercise 4.9)

$$\pi f = E^P\{f \mid \mathcal{I}\}$$

and, because the conditional expectation operator π is well-defined on L_p, as an operator of norm 1 for all p in the range $1 \leq p \leq \infty$. If $f \in L_\infty$, then $\|A_n f\|_\infty \leq \|f\|_\infty$, and by the bounded convergence theorem, for any p satisfying $1 \leq p < \infty$, we have $\|A_n f - \pi f\|_p \to 0$ as $n \to \infty$. Since L_∞ is dense in L_p and $\|A_n\| \leq 1$ in all the L_p-spaces, it is easily seen, by a simple approximation argument, that for each p in $1 \leq p < \infty$ and $f \in L_p$,

$$\lim_{n \to \infty} \|A_n f - f\|_p = 0,$$

proving the mean ergodic theorem in all the L_p-spaces.

We now concentrate on proving almost sure convergence of $A_n f$ to πf for $f \in L_1(P)$. This part is often called the *individual ergodic theorem* or *Birkhoff's theorem*. This will be based on an analogue of Doob's inequality for martingales. First, we will establish an inequality called the *maximal ergodic theorem*.

THEOREM 6.2 (Maximal Ergodic Theorem) *Let $f \in L_1(P)$ and, for $n \geq 1$, let*

$$E_n^0 = \left\{\omega : \sup_{1 \leq j \leq n} [f(\omega) + f(T\omega) + \cdots + f(T^{j-1}\omega)] \geq 0\right\}.$$

Then

$$\int_{E_n^0} f(\omega) dP \geq 0.$$

PROOF: Let

$$h_n(\omega) = \sup_{1 \leq j \leq n} [f(\omega) + f(T\omega) + \cdots + f(T^{j-1}\omega)]$$
$$= f(\omega) + \max(0, h_{n-1}(T\omega)) = f(\omega) + h_{n-1}^+(T\omega)$$

where

$$h_n^+(\omega) = \max(0, h_n(\omega)).$$

On E_n^0, $h_n(\omega) = h_n^+(\omega)$ and therefore

$$f(\omega) = h_n(\omega) - h_{n-1}^+(T\omega) = h_n^+(\omega) - h_{n-1}^+(T\omega).$$

Consequently,

$$\int_{E_n^0} f(\omega) dP = \int_{E_n^0} [h_n^+(\omega) - h_{n-1}^+(T\omega)] dP$$

$$\geq \int_{E_n^0} [h_n^+(\omega) - h_n^+(T\omega)] dP \qquad \text{(because } h_{n-1}^+(\omega) \leq h_n^+(\omega)\text{)}$$

$$* = \int_{E_n^0} h_n^+(\omega) dP - \int_{TE_n^0} h_n^+(\omega) dP \quad \text{(because of invariance of } T)$$
$$\geq 0.$$

The last step follows from the fact that for any integrable function $h(\omega)$, $\int_E h(\omega) dP$ is the largest when we take for E the set $E = \{\omega : h(\omega) \geq 0\}$. $\square$

Now we establish the analogue of Doob's inequality or maximal inequality, or sometimes referred to as the weak-type one-to-one inequality.

LEMMA 6.3 *For any $f \in L_1(P)$ and $\ell > 0$, denoting by E_n the set*
$$E_n = \left\{\omega : \sup_{1 \leq j \leq n} |(A_j f)(\omega)| \geq \ell\right\},$$
we have
$$P[E_n] \leq \frac{1}{\ell} \int_{E_n} |f(\omega)| dP;$$
in particular,
$$P\left[\omega : \sup_{j \geq 1} |(A_j f)(\omega)| \geq \ell\right] \leq \frac{1}{\ell} \int |f(\omega)| dP.$$

PROOF: We can assume without loss of generality that $f \in L_1(P)$ is nonnegative. Apply the lemma to $f - \ell$. If
$$E_n = \left\{\omega : \sup_{1 \leq j \leq n} \frac{[f(\omega) + f(T\omega) + \cdots + f(T^{j-1}\omega)]}{j} \geq \ell\right\},$$
then
$$\int_{E_n} [f(\omega) - \ell] dP \geq 0 \quad \text{or} \quad P[E_n] \leq \frac{1}{\ell} \int_{E_n} f(\omega) dP.$$
We are done. $\square$

Given the lemma, the proof of the almost sure ergodic theorem follows along the same lines as the proof of the almost sure convergence in the martingale context. If $f \in H_0$ it is trivial. For $f = (I - U)g$ with $g \in L_\infty$ it is equally trivial because $\|A_n f\|_\infty \leq \frac{2\|g\|_\infty}{n}$. So the almost sure convergence is valid for $f = f_1 + f_2$ with $f_1 \in H_0$ and $f_2 = (I - U)g$ with $g \in L_\infty$. But such functions are dense in $L_1(P)$. Once we have almost sure convergence for a dense set in $L_1(P)$, the almost sure convergence for every $f \in L_1(P)$ follows by routine approximation using Lemma 6.3; see the proof of Theorem 5.7. $\square$

EXERCISE 6.1. For any bounded linear transformation A on a Hilbert space H, show that the closure of the range of A, i.e., $\overline{\text{Range } A}$, is the orthogonal complement of the null space $\{f : A^* f = 0\}$ where A^* is the adjoint of A.

EXERCISE 6.2. Show that any almost invariant set differs by a set of measure 0 from an invariant set, i.e., if $P(A \triangle T^{-1} A) = 0$, then there is a $B \in \mathcal{F}$ with $P(A \triangle B) = 0$ and $T^{-1} B = B$.

Although the ergodic theorem implies a strong law of large numbers for any stationary sequence of random variables, in particular a sequence of independent identically distributed random variables, it is not quite the end of the story. For the law of large numbers, we need to know that the limit πf is a constant, which will then equal $\int f(\omega) dP$. To claim this, we need to know that the invariant σ-field is trivial or essentially consists of the whole space Ω and the empty set $\varnothing$. An invariant measure P is said to be ergodic for the transformation T if every $A \in \mathcal{I}$; i.e., every invariant set has measure 0 or 1. Then every invariant function is almost surely a constant and $\pi f = E[f \mid \mathcal{I}] = \int f(\omega) dP$.

THEOREM 6.4 *Any product measure is ergodic for the shift.*

PROOF: Let A be an invariant set. Then A can be approximated by sets A_n in the σ-field corresponding to the coordinates from $[-n, n]$. Since A is invariant, $T^{\pm 2n} A_n$ will approximate A just as well. This proves that A actually belongs to the tail σ-field, the remote past as well as the remote future. Now we can use Kolmogorov's zero-one law (Theorem 3.15) to assert that $P(A) = 0$ or 1. □

6.2. Structure of Stationary Measures

Given a space $(\Omega, \mathcal{F})$ and a measurable transformation T with a measurable inverse T^{-1}, we can consider the space $\mathcal{M}$ of all T-invariant probability measures on $(\Omega, \mathcal{F})$. The set $\mathcal{M}$, which may be empty, is easily seen to be a convex set.

EXERCISE 6.3. Let $\Omega = \mathbb{Z}$, the integers, and for $n \in \mathbb{Z}$, let $Tn = n + 1$. Show that $\mathcal{M}$ is empty.

THEOREM 6.5 *A probability measure $P \in \mathcal{M}$ is ergodic if and only if it is an extreme point of $\mathcal{M}$.*

PROOF: A point of a convex set is extreme if it cannot be written as a nontrivial convex combination of two other points from that set. Suppose $P \in \mathcal{M}$ is not extremal. Then P can be written as a nontrivial convex combination of $P_1, P_2 \in \mathcal{M}$, i.e., for some $0 < a < 1$ and $P_1 \neq P_2$, $P = aP_1 + (1-a)P_2$. We claim that such a P cannot be ergodic. If it were, by definition, $P(A) = 0$ or 1 for every $A \in \mathcal{I}$. Since $P(A)$ can be 0 or 1 only when $P_1(A) = P_2(A) = 0$ or $P_1(A) = P_2(A) = 1$, it follows that for every invariant set $A \in \mathcal{I}$, $P_1(A) = P_2(A)$. We now show that if two invariant measures P_1 and P_2 agree on $\mathcal{I}$, they agree on $\mathcal{F}$. Let $f(\omega)$ be any bounded $\mathcal{F}$-measurable function. Consider the function

$$h(\omega) = \lim_{n \to \infty} \frac{1}{n}[f(\omega) + f(T\omega) + \cdots + f(T^{n-1}\omega)]$$

defined on the set E where the limit exists. By the ergodic theorem $P_1(E) = P_2(E) = 1$ and h is $\mathcal{I}$-measurable. Moreover, by the stationarity of P_1 and P_2 and the bounded convergence theorem,

$$E^{P_i}[f(\omega)] = \int_E h(\omega) dP_i \quad \text{for } i = 1, 2.$$

Since $P_1 = P_2$ on $\mathcal{I}$, h is $\mathcal{I}$-measurable, and $P_i(E) = 1$ for $i = 1, 2$, we see that
$$E^{P_1}[f(\omega)] = E^{P_2}[f(\omega)].$$
Since f is arbitrary, this implies that $P_1 = P_2$ on $\mathcal{F}$.

Conversely, if P is not ergodic, then there is an $A \in \mathcal{I}$ with $0 < P(A) < 1$ and we define
$$P_1(E) = \frac{P(A \cap E)}{P(A)}, \quad P_2(E) = \frac{P(A^c \cap E)}{P(A^c)}.$$
Since $A \in \mathcal{I}$, it follows that the P_i are stationary. Moreover, $P = P(A)P_1 + P(A^c)P_2$ and, hence, P is not extremal. □

One of the questions in the theory of convex sets is the existence of sufficiently many extremal points, enough to recover the convex set by taking convex combinations. In particular, one can ask if any point in the convex set can be obtained by taking a weighted average of the extremals. The next theorem answers the question in our context. We will assume that our space $(\Omega, \mathcal{F})$ is nice, i.e., is a complete separable metric space with its Borel sets.

THEOREM 6.6 *For any invariant measure P, there is a probability measure μ_P on the set $\mathcal{M}_e$ of ergodic measures such that*
$$P = \int_{\mathcal{M}_e} Q \, \mu_P(dQ).$$

PROOF: If we denote by P_ω the regular conditional probability distribution of P given $\mathcal{I}$, which exists (see Theorem 4.4) because $(\Omega, \mathcal{F})$ is nice, then
$$P = \int_\Omega P_\omega P(d\omega).$$

We will complete the proof by showing that P_ω is an ergodic stationary probability measure for almost all ω with respect to P. We can then view P_ω as a map $\Omega \to \mathcal{M}_e$ and μ_P will be the image of P under the map. Our integral representation in terms of ergodic measures will just be an immediate consequence of the change-of-variables formula. □

LEMMA 6.7 *For any stationary probability measure P, for almost all ω with respect to P, the regular conditional probability distribution P_ω of P given $\mathcal{I}$ is stationary and ergodic.*

PROOF: Let us first prove stationarity. We need to prove that $P_\omega(A) = P_\omega(TA)$ a.e. We have to negotiate carefully through null sets. Since a measure on the Borel σ-field $\mathcal{F}$ of a complete separable metric space is determined by its values on a countable generating field $\mathcal{F}_0 \subset \mathcal{F}$, it is sufficient to prove that for each fixed $A \in \mathcal{F}_0$, $P_\omega(A) = P_\omega(TA)$ a.e. P. Since P_ω is $\mathcal{I}$-measurable, all we need to show is that, for any $E \in \mathcal{I}$,
$$\int_E P_\omega(A) P(d\omega) = \int_E P_\omega(TA) P(d\omega),$$

or equivalently
$$P(E \cap A) = P(E \cap TA).$$
This is obvious because P is stationary and E is invariant.

We now turn to ergodicity. Again, there is a minefield of null sets to negotiate. It is a simple exercise to check that if, for some stationary measure Q, the ergodic theorem is valid with an almost surely constant limit for the indicator functions $\mathbf{1}_A$ with $A \in \mathcal{F}_0$, then Q is ergodic. This needs to be checked only for a countable collection of sets $\{A\}$. We need, therefore, only to check that any invariant function is constant almost surely with respect to almost all P_ω. Equivalently, for any invariant set E, $P_\omega(E)$ must be shown almost surely to be equal to 0 or 1. But $P_\omega(E) = \chi_E(\omega)$ and is always 0 or 1. This completes the proof. □

EXERCISE 6.4. Show that any two distinct ergodic invariant measures P_1 and P_2 are orthogonal on $\mathcal{I}$; i.e., there is an invariant set E such that $P_1(E) = 1$ and $P_2(E) = 0$.

EXERCISE 6.5. Let $(\Omega, \mathcal{F}) = ([0, 1), \mathcal{B})$ and $Tx = x + a \mod 1$. If a is irrational, there is just one invariant measure P, namely, the uniform distribution on $[0, 1)$. This is seen by Fourier analysis; see Remark 2.2.

$$\int e^{i2n\pi x}\,dP = \int e^{i2n\pi(Tx)}\,dP = \int e^{i2n\pi(x+a)}\,dP = e^{i2n\pi a}\int e^{i2n\pi x}\,dP.$$

If a is irrational, $e^{i2n\pi a} = 1$ if and only if $n = 0$. Therefore,
$$\int e^{i2n\pi x}\,dP = 0 \quad \text{for } n \neq 0,$$
which makes P uniform. Now let $a = \frac{p}{q}$ be rational with $(p, q) = 1$; i.e., p and q are relatively prime. Then, for any x, the discrete distribution with probabilities $\frac{1}{q}$ at the points $\{x, x+a, x+2a, \ldots, x+(q-1)a\}$ is invariant and ergodic. We can denote this distribution by P_x. If we limit x to the interval $0 \leq x < \frac{1}{q}$, then x is uniquely determined by P_x. Complete the example by determining all T-invariant probability distributions on $[0, 1)$ and find the integral representation in terms of the ergodic ones.

6.3. Stationary Markov Processes

Let $\pi(x, dy)$ be a transition probability function on $(X, \mathcal{B})$ where X is a state space and $\mathcal{B}$ is a σ-field of measurable subsets of X. A stochastic process with values in X is a probability measure on the space $(\Omega, \mathcal{F})$, where Ω is the space of sequences $\{x_n : -\infty < n < \infty\}$ with values in X and $\mathcal{F}$ is the product σ-field. The space $(\Omega, \mathcal{F})$ has some natural sub σ-fields. For any two integers $m \leq n$, we have the sub σ-fields, $\mathcal{F}_n^m = \sigma\{x_j : m \leq j \leq n\}$ corresponding to information about the process during the time interval $[m, n]$. In addition, we have $\mathcal{F}_n = \mathcal{F}_n^{-\infty} = \sigma\{x_j : j \leq n\}$ and $\mathcal{F}^m = \mathcal{F}_\infty^m = \sigma\{x_j : j \geq m\}$ that correspond to the past and future. P is a Markov process on $(\Omega, \mathcal{F})$ with transition probability $\pi(\cdot, \cdot)$ if, for every n, $A \in \mathcal{B}$ and P-almost all ω,
$$P\{x_{n+1} \in A \mid \mathcal{F}_n\} = \pi(x_n, A).$$

REMARK 6.1. Given a π, it is not always true that P exists. A simple but illuminating example is to take $X = \{0, 1, \ldots, n, \ldots\}$ to be the nonnegative integers and define $\pi(x, x+1) = 1$, and all the process does is move one step to the right every time. Such a process, if it had started a long time back, would be found nowhere today! So it does not exist. On the other hand, if we take X to be the set of all integers, then P is seen to exist. In fact, there are lots of them. What is true, however, is that given any initial distribution μ and initial time m, there exists a unique process P on $(\Omega, \mathcal{F}^m)$, i.e., defined on the future σ-field from time m on, that is Markov with transition probability π and satisfies $P\{x_m \in A\} = \mu(A)$ for all $A \in \mathcal{B}$.

The shift T acts naturally as a measurable invertible map of the product space Ω into itself, and the notion of a stationary process makes sense. The following theorem connects stationarity and the Markov property:

THEOREM 6.8 *Let the transition probability π be given. Let P be a stationary Markov process with transition probability π. Then the one-dimensional marginal distribution μ, which is independent of time because of stationarity and given by*

$$\mu(A) = P\{x_n \in A\},$$

is π-invariant in the sense that

$$\mu(A) = \int \pi(x, A)\mu(dx)$$

for every set $A \in \mathcal{B}$. Conversely, given such a μ, there is a unique stationary Markov process P with marginals μ and transition probability π.

EXERCISE 6.6. Prove the above theorem; use Remark 4.7.

EXERCISE 6.7. If P is a stationary Markov process on a countable state space with transition probability π and invariant marginal distribution μ, show that the time-reversal map that maps $\{x_n\}$ to $\{x_{-n}\}$ takes P to another stationary Markov process Q, and express the transition probability $\hat{\pi}$ of Q as explicitly as you can in terms of π and μ.

EXERCISE 6.8. If μ is an invariant measure for π, show that the conditional expectation map $\Pi : f(\cdot) \to \int f(y)\pi(\cdot, dy)$ induces a contraction in $L_p(\mu)$ for any $p \in [1, \infty]$. We say that a Markov process is reversible if the time-reversed process Q of the previous example coincides with P. Show that P corresponding to π and μ is reversible if and only if the corresponding Π in $L_2(\mu)$ is self-adjoint or symmetric.

Since a given transition probability π may in general have several invariant measures μ, there will be several stationary Markov processes with transition probability π. Let $\widetilde{\mathcal{M}}$ be the set of invariant probability measures for the transition probabilities $\pi(x, dy)$, i.e.,

$$\widetilde{\mathcal{M}} = \left\{\mu : \mu(A) = \int_X \pi(x, A)d\mu(x) \text{ for all } A \in \mathcal{B}\right\}.$$

6.3. STATIONARY MARKOV PROCESSES

$\widetilde{\mathcal{M}}$ is a convex set of probability measures, and we denote by $\widetilde{\mathcal{M}}_e$ its (possibly empty) set of extremals. For each $\mu \in \widetilde{\mathcal{M}}$, we have the corresponding stationary Markov process P_μ, and the map $\mu \to P_\mu$ is clearly linear. If we want P_μ to be an ergodic stationary process, then it must be an extremal in the space of all stationary processes. The extremality of $\mu \in \widetilde{\mathcal{M}}$ is, therefore, a necessary condition for P_μ to be ergodic. That it is also sufficient is a little bit of a surprise. The following theorem is the key step in the proof. The remaining part is routine.

THEOREM 6.9 *Let μ be an invariant measure for π and $P = P_\mu$ the corresponding stationary Markov process. Let $\mathcal{I}$ be the σ-field of shift-invariant subsets on Ω. To within sets of P measure 0, $\mathcal{I} \subset \mathcal{F}_0^0$.*

PROOF: This theorem describes completely the structure of nontrivial sets in the σ-field $\mathcal{I}$ of invariant sets for a stationary Markov process with transition probability π and marginal distribution μ. Suppose that the state space can be partitioned nontrivially, i.e., with $0 < \mu(A) < 1$, into two sets A and A^c that satisfy $\pi(x, A) = 1$ a.e. μ on A and $\pi(x, A^c) = 1$ a.e. μ on A^c. Then the event

$$E = \{\omega : x_n \in A \text{ for all } n \in \mathbb{Z}\}$$

provides a nontrivial set in $\mathcal{I}$. The theorem asserts the converse. The proof depends on the fact that an invariant set E is in the remote past $\mathcal{F}_{-\infty}^{-\infty} = \bigcap_n \mathcal{F}_n^{-\infty}$ as well as in the remote future $\mathcal{F}_\infty^\infty = \bigcap_m \mathcal{F}_\infty^m$. See the proof of Theorem 6.4. For a Markov process the past and the future are conditionally independent given the present; see Theorem 4.9. This implies that

$$P[E \mid \mathcal{F}_0^0] = P[E \cap E \mid \mathcal{F}_0^0] = P[E \mid \mathcal{F}_0^0] P[E \mid \mathcal{F}_0^0]$$

and must therefore equal either 0 or 1. This, in turn, means that corresponding to any invariant set $E \in \mathcal{I}$, there exists $A \subset X$ that belongs to $\mathcal{B}$ such that $E = \{\omega : x_n \in A \text{ for all } n \in \mathbb{Z}\}$ up to a set of P measure 0. If the Markov process starts from A or A^c, it does not ever leave it. That means $0 < \mu(A) < 1$ and

$$\pi(x, A^c) = 0 \text{ for } \mu \text{ a.e. } x \in A \quad \text{and} \quad \pi(x, A) = 0 \text{ for } \mu \text{ a.e. } x \in A^c.$$

□

REMARK 6.2. One way to generate Markov processes with multiple invariant measures is to start with two Markov processes with transition probabilities $\pi_i(x_i, dy_i)$ on X_i and invariant measures μ_i, and consider $X = X_1 \cup X_2$. Define

$$\pi(x, A) = \begin{cases} \pi_1(x, A \cap X_1) & \text{if } x \in X_1 \\ \pi_2(x, A \cap X_2) & \text{if } x \in X_2. \end{cases}$$

Then any one of the two processes can be going on depending on which world we are in. Both μ_1 and μ_2 are invariant measures. We have combined two distinct possibilities into one. What we have shown is that when we have multiple invariant measures, they essentially arise in this manner.

REMARK 6.3. We can therefore look at the convex set of measures μ that are π-invariant, i.e., $\mu\pi = \mu$. The extremals of this convex set are precisely the ones that correspond to ergodic stationary processes, and they are called *ergodic* or *extremal invariant measures*. If the set of invariant probability measures is nonempty for some π, then there are enough extremals to recover any arbitrary invariant measure as an integral or weighted average of extremal ones.

EXERCISE 6.9. Show that any two distinct extremal invariant measures μ_1 and μ_2 for the same π are orthogonal on $\mathcal{B}$.

EXERCISE 6.10. Consider the operator Π on the $L_p(\mu)$-spaces corresponding to a given invariant measure. The dimension of the eigenspace $f : \Pi f = f$ that corresponds to the eigenvalue 1 determines the extremality of μ. Clarify this statement.

EXERCISE 6.11. Let P_x be the Markov process with stationary transition probability $\pi(x, dy)$ starting at time 0 from $x \in X$. Let f be a bounded measurable function on X. Then, for almost all x with respect to any extemal invariant measure ν,
$$\lim_{n \to \infty} \frac{1}{n}[f(x_1) + f(x_2) + \cdots + f(x_n)] = \int f(y)\nu(dy)$$
for almost all ω with respect to P_x.

EXERCISE 6.12. We saw in the earlier section that any stationary process is an integral over stationary ergodic processes. If we represent a stationary Markov process P_μ as the integral
$$P_\mu = \int R Q(dR)$$
over stationary ergodic processes, show that the integral really involves only stationary Markov processes with the same transition probability π, so that the integral is really of the form
$$P_\mu = \int_{\widetilde{\mathcal{M}}_e} P_\nu Q(d\nu) \quad \text{or equivalently} \quad \mu = \int_{\widetilde{\mathcal{M}}_e} \nu Q(d\nu).$$

EXERCISE 6.13. If there is a reference measure α such that $\pi(x, dy)$ has a density $p(x, y)$ with respect to α for every x, then show that any invariant measure μ is absolutely continuous with respect to α.

The question of when there is at most one invariant measure for the Markov process with transition probability π is a difficult one. If we have a density $p(x, y)$ with respect to a reference measure α and if, for each x, $p(x, y) > 0$ for almost all y with respect to α, then there can be at most one invariant measure. We saw already that any invariant measure has a density with respect to α. If there are at least two invariant measures, then there are at least two ergodic ones that are orthogonal. If we denote by f_1 and f_2 their densities with respect to α, by orthogonality we know that they are supported on disjoint invariant sets, A_1 and A_2; in particular, $p(x, y) = 0$ for almost all x on A_1 in the support of f_1 and almost all y in A_2 with respect to α. By our positivity assumption we must have $\alpha(A_2) = 0$, which is a contradiction.

6.4. Mixing Properties of Markov Processes

One of the questions that is important in the theory of Markov processes is the rapidity with which the memory of the initial state is lost. There is no unique way of assessing it, and, depending on the circumstances, this could happen in many different ways at many different rates. Let $\pi^{(n)}(x, dy)$ be the n-step transition probability. The issue is how the measures $\pi^{(n)}(x, dy)$ depend less and less on x as $n \to \infty$. Suppose we measure this dependence by

$$\rho_n = \sup_{x,y \in X} \sup_{A \in \mathcal{B}} \left| \pi^{(n)}(x, A) - \pi^{(n)}(y, A) \right|;$$

then the following is true:

THEOREM 6.10 *Either $\rho_n \equiv 1$ for all $n \geq 1$ or $\rho_n \leq C\theta^n$ for some $0 \leq \theta < 1$.*

PROOF: From the Chapman-Kolmogorov equations

$$\pi^{(n+m)}(x, A) - \pi^{(n+m)}(y, A) = \int \pi^{(m)}(z, A)[\pi^{(n)}(x, dz) - \pi^{(n)}(y, dz)].$$

If $f(\cdot)$ is a function with $|f(x) - f(y)| \leq C$ for all x and y, and $\mu = \mu_1 - \mu_2$ is the difference of two probability measures with $\|\mu\| = \sup_A |\mu(A)| \leq \delta$, then it is elementary to estimate, using $\mu(X) = 0$,

$$\left| \int f \, d\mu \right| = \inf_c \left| \int (f - c) d\mu \right| \leq 2 \inf_c \left\{ \sup_x |f(x) - c| \right\} \|\mu\| \leq 2 \frac{C}{2} \delta = C\delta.$$

It follows that the sequence ρ_n is submultiplicative, i.e.,

$$\rho_{m+n} \leq \rho_m \rho_n.$$

Our theorem follows from this property. As soon as some $\rho_k = a < 1$, we have

$$\rho_n \leq [\rho_k]^{\lfloor \frac{n}{k} \rfloor} \leq C\theta^n$$

with $\theta = a^{1/k}$. □

Although this is an easy theorem, it can be applied in some contexts.

REMARK 6.4. If $\pi(x, dy)$ has density $p(x, y)$ with respect to some reference measure α and $p(x, y) \geq q(y) \geq 0$ for all y with $\int q(y) d\alpha \geq \delta > 0$, then it is elementary to show that $\rho_1 \leq (1 - \delta)$.

REMARK 6.5. If $\rho_n \to 0$, we can estimate

$$|\pi^{(n)}(x, A) - \pi^{(n+m)}(x, A)| = \left| \int [\pi^{(n)}(x, A) - \pi^{(n)}(y, A)] \pi^{(m)}(x, dy) \right| \leq \rho_n$$

and conclude from the estimate that

$$\lim_{n \to \infty} \pi^{(n)}(x, A) = \mu(A)$$

exists. μ is seen to be an invariant probability measure.

REMARK 6.6. In this context the invariant measure is unique. If β is another invariant measure, because

$$\beta(A) = \int \pi^{(n)}(x, A)\beta(dy)$$

for every $n \geq 1$,

$$\beta(A) = \lim_{n \to \infty} \int \pi^{(n)}(y, A)\beta(dy) = \mu(A).$$

REMARK 6.7. The stationary process P_μ has the property that if $E \in \mathcal{F}_m^{-\infty}$ and $F \in \mathcal{F}_\infty^n$ with a gap of $k = n - m > 0$, then

$$P_\mu[E \cap F] = \int_E \int_X \pi^{(k)}(x_m(\omega), dx) P_x(T^{-n}F) P_\mu(d\omega)$$

$$P_\mu[E] P_\mu[F] = \int_E \int_X \mu(dx) P_x(T^{-n}F) P_\mu(d\omega)$$

$$P_\mu[E \cap F] - P_\mu[E] P_\mu[F] =$$

$$\int_E \int_X P_x(T^{-n}F) [\pi^{(k)}(x_m(\omega), dx) - \mu(dx)] P_\mu(d\omega),$$

from which it follows that

$$|P_\mu[E \cap F] - P_\mu[E] P_\mu[F]| \leq \rho_k P_\mu(E),$$

proving an asymptotic independence property for P_μ.

There are situations in which we know that an invariant probability measure μ exists for π and we wish to establish that $\pi^{(n)}(x, A)$ converges to $\mu(A)$ uniformly in A for each $x \in X$ but not necessarily uniformly over the starting points x. Uniformity in the starting point is very special. We will illustrate this by an example.

EXAMPLE 6.1. The Ornstein-Uhlenbeck process is a Markov chain on the state space $X = \mathbb{R}$, the real line, with transition probability $\pi(x, dy)$ given by a Gaussian distribution with mean ρx and variance σ^2. It has a density $p(x, y)$ with respect to the Lebesgue measure, so that $\pi(x, A) = \int_A p(x, y) dy$.

$$p(x, y) = \frac{1}{\sqrt{2\pi}\sigma} \exp\left[-\frac{(y - \rho x)^2}{2\sigma^2}\right].$$

It arises from the "autoregressive" representation

$$x_{n+1} = \rho x_n + \sigma \xi_{n+1},$$

where $\xi_1, \xi_2, \ldots, \xi_n \ldots$ are independent standard Gaussians. The characteristic function of any invariant measure $\phi(t)$ satisfies, for every $n \geq 1$,

$$\phi(t) = \phi(\rho t) \exp\left[-\frac{\sigma^2 t^2}{2}\right] = \phi(\rho^n t) \exp\left[-\frac{(\sum_{j=0}^{n-1} \rho^{2j})\sigma^2 t^2}{2}\right]$$

by induction on n. Therefore,

$$|\phi(t)| \leq \exp\left[-\frac{(\sum_{j=0}^{n-1}\rho^{2j})\sigma^2 t^2}{2}\right],$$

and this cannot be a characteristic function unless $|\rho| < 1$ (otherwise by letting $n \to \infty$ we see that $\phi(t) = 0$ for $t \neq 0$ and therefore discontinuous at $t = 0$). If $|\rho| < 1$, by letting $n \to \infty$ and observing that $\phi(\rho^n t) \to \phi(0) = 1$,

$$\phi(t) = \exp\left[-\frac{\sigma^2 t^2}{2(1-\rho^2)}\right].$$

The only possible invariant measure is the Gaussian with mean 0 and variance $\frac{\sigma^2}{(1-\rho^2)}$. One can verify that this Gaussian is in fact, an invariant measure. If $|\rho| < 1$, a direct computation shows that $\pi^{(n)}(x, dy)$ is a Gaussian with mean $\rho^n x$ and variance $\sigma_n^2 = \sum_{j=0}^{n-1} \rho^{2j}\sigma^2 \to (1-\rho^2)^{-1}\sigma^2$ as $n \to \infty$. Clearly, there is uniform convergence only over bounded sets of starting points x. This is typical.

6.5. Central Limit Theorem for Martingales

If $\{\xi_n\}$ is an ergodic stationary sequence of random variables with mean 0, then we know from the ergodic theorem that the mean $\frac{\xi_1+\xi_2+\cdots+\xi_n}{n}$ converges to zero almost surely. We want to develop some methods for proving the central limit theorem, i.e., the convergence of the distribution of $\frac{\xi_1+\xi_2+\cdots+\xi_n}{\sqrt{n}}$ to some Gaussian distribution with mean 0 and variance σ^2. Under the best of situations, since the covariance $\rho_k = E[X_n X_{n+k}]$ may not be 0 for all $k \neq 0$, if we assume that $\sum_{-\infty < j < \infty} |\rho_j| < \infty$, we get

$$\sigma^2 = \lim_{n \to \infty} \frac{1}{n} E[(\xi_1 + \xi_2 + \cdots + \xi_n)^2] = \lim_{n \to \infty} \sum_{|j| \leq n} \left(1 - \frac{|j|}{n}\right)\rho_j$$

$$= \sum_{-\infty < j < \infty} \rho_j = \rho_0 + 2\sum_{j=1}^{\infty} \rho_j.$$

The standard central limit theorem with $\sqrt{n}$ scaling is not likely to work if the covariances do not decay rapidly enough to be summable. When the covariances $\{\rho_k\}$ are all 0 for $k \neq 0$, the variance calculation yields $\sigma^2 = \rho_0$ just as in the independent case, but there is no guarantee that the central limit theorem is valid.

A special situation is when $\{\xi_j\}$ are square-integrable martingale differences. With the usual notation for the σ-fields $\mathcal{F}_n^m$ for $m \leq n$ (remember that m can be $-\infty$ while n can be $+\infty$), we assume that

$$E\{\xi_n \mid \mathcal{F}_{n-1}\} = 0 \quad \text{a.e.},$$

and in this case by conditioning we see that $\rho_k = 0$ for $k \neq 0$. It is a useful and important observation that in this context the central limit theorem always holds. The distribution of $Z_n = \frac{\xi_1+\xi_2+\cdots+\xi_n}{\sqrt{n}}$ converges to the normal distribution with mean 0 and variance $\sigma^2 = \rho_0$. The proof is a fairly simple modification of the usual proof of the central limit theorem. Let us define

$$\psi(n, j, t) = \exp\left[\frac{\sigma^2 t^2 j}{2n}\right] E\left\{\exp\left[it\frac{\xi_1 + \xi_2 + \cdots + \xi_j}{\sqrt{n}}\right]\right\}$$

and write

$$\psi(n, n, t) - 1 = \sum_{j=1}^{n} [\psi(n, j, t) - \psi(n, j-1, t)],$$

leaving us with the estimation of

$$\Delta(n, t) = \left|\sum_{j=1}^{n} [\psi(n, j, t) - \psi(n, j-1, t)]\right|.$$

THEOREM 6.11 *For an ergodic stationary sequence $\{\xi_j\}$ of square-integrable martingale differences, the central limit theorem is always valid.*

PROOF: We let $S_j = \xi_1 + \xi_2 + \cdots + \xi_j$ and calculate

$$[\psi(n, j, t) - \psi(n, j-1, t)] =$$

$$\exp\left[\frac{\sigma^2 t^2 j}{2n}\right] E\left\{\exp\left[it\frac{S_{j-1}}{\sqrt{n}}\right]\left[\exp\left[it\frac{\xi_j}{\sqrt{n}}\right] - \exp\left[-\frac{\sigma^2 t^2}{2n}\right]\right]\right\}.$$

We can replace it with

$$\theta(n, j, t) = \exp\left[\frac{\sigma^2 t^2 j}{2n}\right] E\left\{\exp\left[it\frac{S_{j-1}}{\sqrt{n}}\right]\left[\frac{(\sigma^2 - \xi_j^2)t^2}{2n}\right]\right\}$$

because the error can be controlled by Taylor's expansion. In fact, if we use the martingale difference property to kill the linear term, we can bound the difference, in an arbitrary finite interval $|t| \leq T$, by

$$|[\psi(n, j, t) - \psi(n, j-1, t)] - \theta(n, j, t)|$$

$$\leq C_T E\left\{\left|\exp\left[it\frac{\xi_j}{\sqrt{n}}\right] - 1 - it\frac{\xi_j}{\sqrt{n}} + \frac{t^2 \xi_j^2}{2n}\right|\right\}$$

$$+ C_T \left|\exp\left[-\frac{\sigma^2 t^2}{2n}\right] - 1 + \frac{\sigma^2 t^2}{2n}\right|,$$

where C_T is a constant that depends only on T. The right-hand side is independent of j because of stationarity. By Taylor expansions in the variable $t/\sqrt{n}$ of each of the two terms on the right, it is easily seen that

$$\sup_{\substack{|t| \leq T \\ 1 \leq j \leq n}} |[\psi(n, j, t) - \psi(n, j-1, t)] - \theta(n, j, t)| = o\left(\frac{1}{n}\right).$$

Therefore,

$$\sup_{|t| \leq T} \sum_{j=1}^{n} |[\psi(n, j, t) - \psi(n, j-1, t)] - \theta(n, j, t)| = n \cdot o\left(\frac{1}{n}\right) \to 0.$$

We now concentrate on estimating $|\sum_{j=1}^n \theta(n, j, t)|$. We pick an integer k which will be large but fixed. We divide $[1, n]$ into blocks of size k with perhaps an incomplete block at the end. We will now replace $\theta(n, j, t)$ by

$$\theta_k(n, j, t) = \exp\left[\frac{\sigma^2 t^2 kr}{2n}\right] E\left\{\exp\left[it\frac{S_{kr}}{\sqrt{n}}\right]\left[\frac{(\sigma^2 - \xi_j^2)t^2}{2n}\right]\right\}$$

for $kr + 1 \leq j \leq k(r+1)$ and $r \geq 0$.

Using stationarity it is easy to estimate, for $r \leq \frac{n}{k}$,

$$\left|\sum_{j=kr+1}^{k(r+1)} \theta_k(n, j, t)\right| \leq C(t)\frac{1}{n} E\left\{\left|\sum_{j=kr+1}^{k(r+1)} (\sigma^2 - \xi_j^2)\right|\right\} = C(t)\frac{k}{n}\delta(k),$$

where $\delta(k) \to 0$ as $k \to \infty$ by the L_1-ergodic theorem. After all, $\{\xi_j^2\}$ is a stationary sequence with mean σ^2 and the ergodic theorem applies. The above estimate is uniform in r, the incomplete block at the end causes no problem, and there are approximately $\frac{n}{k}$ blocks. We can now conclude that

$$\left|\sum_{j=1}^{n} \theta_k(n, j, t)\right| \leq C(t)\delta(k).$$

On the other hand, by stationarity,

$$\sum_{j=1}^{n} |\theta_k(n, j, t) - \theta(n, j, t)|$$

$$\leq n \sup_{1 \leq j \leq n} |\theta_k(n, j, t) - \theta(n, j, t)|$$

$$\leq C(t) \sup_{1 \leq j \leq k} E\left\{\left|\exp\left[\frac{\sigma^2 t^2 j}{2n}\right]\exp\left[it\frac{S_{j-1}}{\sqrt{n}}\right] - 1\right||\sigma^2 - \xi_j^2|\right\},$$

and it is elementary to show by the dominated convergence theorem that the right-hand side tends to 0 as $n \to \infty$ for each finite k.

This concludes the proof of the theorem. $\square$

One may think that the assumption that $\{\xi_n\}$ is a martingale difference is too restrictive to be useful. Let $\{X_n\}$ be any stationary process with zero mean. We can often succeed in writing $X_n = \xi_{n+1} + \eta_{n+1}$ where ξ_n is a martingale difference and η_n is negligible in the sense that $E[(\sum_{j=1}^n \eta_j)^2] = o(n)$. Then the central limit theorem for $\{X_n\}$ can be deduced from that of $\{\xi_n\}$. A cheap way to prove $E[(\sum_{j=1}^n \eta_j)^2] = o(n)$ is to establish that $\eta_n = Z_n - Z_{n+1}$ for some stationary square-integrable sequence $\{Z_n\}$. Then $\sum_{j=1}^n \eta_j$ telescopes and the needed estimate is obvious. Here is a way to construct Z_n from X_n so that $X_n + (Z_{n+1} - Z_n)$ is a martingale difference.

Let us define

$$Z_n = \sum_{j=0}^{\infty} E\{X_{n+j} \mid \mathcal{F}_n\}.$$

There is no guarantee that the series converges, but we can always hope. After all, if the memory is weak, prediction j steps ahead should be futile if j is large. Therefore, if X_{n+j} is becoming independent of $\mathcal{F}_n$ as j gets large, one would expect $E\{X_{n+j} \mid \mathcal{F}_n\}$ to approach $E[X_{n+j}]$, which is assumed to be 0. By stationarity n plays no role. If Z_0 can be defined, the shift operator T can be used to define $Z_n(\omega) = Z_0(T^n\omega)$. Let us assume that the $\{Z_n\}$ exist and are square-integrable. Then

$$Z_n = E\{Z_{n+1} \mid \mathcal{F}_n\} + X_n$$

or, equivalently,

$$X_n = Z_n - E\{Z_{n+1} \mid \mathcal{F}_n\}$$
$$= [Z_n - Z_{n+1}] + [Z_{n+1} - E\{Z_{n+1} \mid \mathcal{F}_n\}] = \eta_{n+1} + \xi_{n+1},$$

where $\eta_{n+1} = Z_n - Z_{n+1}$ and $\xi_{n+1} = Z_{n+1} - E\{Z_{n+1} \mid \mathcal{F}_n\}$. It is easy to see that $E[\xi_{n+1} \mid \mathcal{F}_n] = 0$.

For a stationary ergodic Markov process $\{X_n\}$ on state space $(X, \mathcal{B})$, with transition probability $\pi(x, dy)$ and invariant measure μ, we can prove the central limit theorem by this method. Let $Y_j = f(X_j)$. Using the Markov property, we can calculate

$$Z_0 = \sum_{j=0}^{\infty} E[f(X_j) \mid \mathcal{F}_0] = \sum_{j=0}^{\infty} [\Pi^j f](X_0) = \left[[I - \Pi]^{-1} f\right](X_0).$$

If the equation $[I - \Pi]U = f$ can be solved with $U \in L_2(\mu)$, then

$$\xi_{n+1} = U(X_{n+1}) - U(X_n) + f(X_n)$$

is a martingale difference and we have a central limit theorem for $\frac{\sum_{j=1}^n f(X_j)}{\sqrt{n}}$ with variance given by

$$\sigma^2 = E^{P_\mu}\{[\xi_0]^2\} = E^{P_\mu}\{[U(X_1) - U(X_0) + f(X_0)]^2\}.$$

EXERCISE 6.14. Let us consider a two-state Markov chain with states $\{1, 2\}$. Let the transition probabilities be given by $\pi(1, 1) = \pi(2, 2) = p$ and $\pi(1, 2) = \pi(2, 1) = q$ with $0 < p, q < 1$ and $p + q = 1$. The invariant measure is given by $\mu(1) = \mu(2) = \frac{1}{2}$ for all values of p. Consider the random variable $S_n = A_n - B_n$, where A_n and B_n are, respectively, the number of visits to the states 1 and 2 during the first n steps. Prove a central limit theorem for $S_n/\sqrt{n}$ and calculate the limiting variance as a function $\sigma^2(p)$ of p. How does $\sigma^2(p)$ behave as $p \to 0$ or 1? Can you explain it? What is the value of $\sigma^2(\frac{1}{2})$? Could you have guessed it?

EXERCISE 6.15. Consider a random walk on the nonnegative integers with

$$\pi(x, y) = \begin{cases} \frac{1}{2} & \text{for all } x = y \geq 0 \\ \frac{1-\delta}{4} & \text{for } y = x + 1, x \geq 1 \\ \frac{1+\delta}{4} & \text{for } y = x - 1, x \geq 1 \\ \frac{1}{2} & \text{for } x = 0, y = 1. \end{cases}$$

Prove that the chain is positive recurrent and find the invariant measure $\mu(x)$ explicitly. If $f(x)$ is a function on $x \geq 0$ with compact support, solve explicitly the equation $[I - \Pi]U = f$. Show that either U grows exponentially at infinity or is a constant for large x. Show that it is a constant if and only if $\sum_x f(x)\mu(x) = 0$. What can you say about the central limit theorem for $\sum_{j=0}^n f(X_j)$ for such functions f?

6.6. Stationary Gaussian Processes

Considering the importance of Gaussian distributions in probability theory, it is only natural to study stationary Gaussian processes, i.e., stationary processes $\{X_n\}$ that have Gaussian distributions as their finite-dimensional joint distributions. Since a joint Gaussian distribution is determined by its means and covariances, we need only specify $E[X_n]$ and $\text{Cov}(X_n, X_m) = E[X_n X_m] - E[X_n]E[X_m]$. Recall that the joint density on $\mathbb{R}^N$ of N Gaussian random variables with mean $\mu = \{\mu_i\}$ and covariance $C = \{\rho_{i,j}\}$ is given by

$$p(y) = \left[\frac{1}{\sqrt{2\pi}}\right]^N \frac{1}{\sqrt{\text{Det } C}} \exp\left[-\frac{1}{2}\langle (y - \mu), C^{-1}(y - \mu)\rangle\right].$$

Here m is the vector of means and C^{-1} is the inverse of the positive definite covariance matrix C. If C is only positive semidefinite, the Gaussian distribution lives on a lower-dimensional hyperplane and is singular. By stationarity $E[X_n] = c$ is independent of n and $\text{Cov}(X_n, X_m) = \rho_{n-m}$ can depend only on the difference $n - m$. By symmetry $\rho_k = \rho_{-k}$. Because the covariance matrix is always positive semidefinite, the sequence ρ_k has the positive definiteness property

$$\sum_{k,j=1}^n \rho_{j-k} z_j \bar{z}_k \geq 0$$

for all choices of n and complex numbers $z_1, z_2, \ldots, z_n$. By Bochner's theorem (see Theorem 2.2) there exists a nonnegative measure μ on the circle that is thought of as $S = [0, 2\pi]$ with endpoints identified such that

$$\rho_k = \int_0^{2\pi} \exp[ik\theta]d\mu(\theta),$$

and because of the symmetry of ρ_k, μ is symmetric as well with respect to $\theta \to 2\pi - \theta$. It is convenient to assume that $c = 0$. One can always add it back.

Given a Gaussian process, it is natural to carry out linear operations that will leave the Gaussian character unchanged. Rather than working with the σ-fields $\mathcal{F}_n^m$, we will work with the linear subspaces $\mathcal{H}_n^m$ spanned by $\{X_j : m \leq j \leq n\}$ and the infinite spans $\mathcal{H}_n = \bigvee_{m \leq n} \mathcal{H}_n^m$ and $\mathcal{H}^m = \bigvee_{n \geq m} \mathcal{H}_n^m$ that are considered as linear subspaces of the Hilbert space

$$\mathcal{H} = \bigvee_{\substack{m,n \\ n \geq m}} \mathcal{H}_n^m,$$

which lies inside $L_2(P)$. But $\mathcal{H}$ is a small part of $L_2(P)$, consisting of only linear functions of $\{X_j\}$. The analogue of Kolmogorov's tail σ-field are the subspaces

$\bigwedge_m \mathcal{H}^m$ and $\bigwedge_n \mathcal{H}_n$ that are denoted by $\mathcal{H}^\infty$ and $\mathcal{H}_{-\infty}$. The analogue of Kolmogorov's zero-one law would be that these subspaces are trivial, having in them only the zero function. The symmetry in ρ_k implies that the processes $\{X_n\}$ and $\{X_{-n}\}$ have the same underlying distributions so that both tails behave identically. A stationary Gaussian process $\{X_n\}$ with mean 0 is said to be purely nondeterministic if the tail subspaces are trivial.

In finite-dimensional theory a covariance matrix can be diagonalized or, better still, written in the special form T^*T, which gives a linear representation of the Gaussian random variables in terms of canonical or independent standard Gaussian random variables. The point to note is that if X is standard Gaussian with mean 0 and covariance $I = \{\delta_{i,j}\}$, then for any linear transformation T, $Y = TX$ is again Gaussian with mean 0 and covariance $C = TT^*$; in other words, if

$$Y_i = \sum_k t_{i,k} X_k \quad \text{then} \quad C_{i,j} = \sum_k t_{i,k} t_{j,k} \,.$$

In fact, for any C we can find a T which is upper or lower triangular, i.e., $t_{i,k} = 0$ for $i > k$ or $i < k$. If the indices correspond to time, this can be interpreted as a causal representation in terms of current and future or past variables only.

The following questions have simple answers:

Question 1. When does a Gaussian process have a moving-average representation in terms of independent Gaussians, i.e., a representation of the form

$$X_n = \sum_{m=-\infty}^{\infty} a_{n-m} \xi_m \quad \text{with} \quad \sum_{n=-\infty}^{\infty} a_n^2 < \infty$$

in terms of i.i.d. Gaussians $\{\xi_k\}$ with mean 0 and variance 1?

If we have such a representation, then the covariance ρ_k is easily calculated as the convolution

$$\rho_k = \sum_j a_j a_{j+k} = [a * \bar{a}](k)\,,$$

and that will make $\{\rho_k\}$ the Fourier coefficients of the function

$$f = \left|\sum_j a_j e^{ij\theta}\right|^2,$$

which is the square of a function in $L_2(S)$. In other words, the spectral measure μ will be absolutely continuous with a density f with respect to the normalized Lebesgue measure $\frac{d\theta}{2\pi}$. Conversely, if we have a μ with a density f, its square root will be a function in L_2 and therefore have Fourier coefficients a_n in l_2, and a moving-average representation holds in terms of i.i.d. random variables with these weights.

Question 2. When does a Gaussian process have a representation that is causal, i.e., of the form

$$X_n = \sum_{j \geq 0} a_j \xi_{n-j} \quad \text{with} \quad \sum_{j \geq 0} a_j^2 < \infty ?$$

If we do have a causal representation, then the remote past of the $\{X_k\}$ process is clearly part of the remote past of the $\{\xi_k\}$ process. By Kolmogorov's zero-one law, the remote past for independent Gaussians is trivial, and a causal representation is therefore possible for $\{X_k\}$ only if its remote past is trivial. The converse is true as well. The subspace $\mathcal{H}_n$ is spanned by $\mathcal{H}_{n-1}$ and X_n. Therefore, either $\mathcal{H}_n = \mathcal{H}_{n-1}$ or $\mathcal{H}_{n-1}$ has codimension 1 in $\mathcal{H}_n$. In the former case, by stationarity $\mathcal{H}_n = \mathcal{H}_{n-1}$ for every n. This, in turn, implies $\mathcal{H}_{-\infty} = \mathcal{H} = \mathcal{H}^\infty$. Assuming that the process is not identically zero, i.e., $\rho_0 = \mu(S) > 0$, this makes the remote past or future the whole thing and definitely nontrivial. So we may assume that $\mathcal{H}_n = \mathcal{H}_{n-1} \oplus \mathbf{e_n}$ where $\mathbf{e_n}$ is a one-dimensional subspace spanned by a unit vector ξ_n. Since all our random variables are linear combinations of a Gaussian collection, they all have Gaussian distributions. We have the shift operator U satisfying $UX_n = X_{n+1}$, and we can assume without loss of generality that $U\xi_n = \xi_{n+1}$ for every n. If we start with X_0 in our Hilbert space

$$X_0 = a_0 \xi_0 + R_{-1}$$

with $R_1 \in \mathcal{H}_{n-1}$. We can continue and write

$$R_{-1} = a_1 \xi_{-1} + R_{-2},$$

and so on. We will then have, for every n,

$$X_0 = a_0 \xi_0 + a_1 \xi_{-1} + \cdots + a_n \xi_{-n} + R_{-(n+1)}$$

with $R_{-(n+1)} \in \mathcal{H}_{-(n+1)}$. Since $\bigwedge_n \mathcal{H}_{-n} = \{0\}$, we conclude that the expansion

$$X_0 = \sum_{j=0}^{\infty} a_j \xi_{-j}$$

is valid.

Question 3. What are the conditions on the spectral density f in order that the process may admit a causal representation?

From our answer to question 1, we know that we have to solve the following analytical problem. Given the spectral measure μ with a nonnegative density $f \in L_1(S)$, when can we write $f = |g|^2$ for some $g \in L_2(S)$ that admits a Fourier representation $g = \sum_{j \geq 0} a_j e^{ij\theta}$ involving only positive frequencies? This has the following neat solution, which is far from obvious:

THEOREM 6.12 *The process determined by the spectral density f admits a causal representation if and only if $f(\theta)$ satisfies*

$$\int_S \log f(\theta) d\theta > -\infty.$$

REMARK 6.8. Notice that the condition basically prevents f from vanishing on a set of positive measure or having very flat zeros.

PROOF: The proof will use methods from the theory of functions of a complex variable. Define

$$g(\theta) = \sum_{n \geq 0} c_n \exp[in\theta]$$

as the Fourier series of some $g \in L_2(S)$. Assume $c_n \neq 0$ for some $n > 0$. In fact, we can assume without loss of generality that $c_0 \neq 0$ by removing a suitable factor of $e^{ik\theta}$ which will not affect $|g(\theta)|$. Then we will show that

$$\frac{1}{2\pi} \int_S \log |g(\theta)| d\theta \geq \log |c_0|.$$

Consider the function
$$G(z) = \sum_{n \geq 0} c_n z^n$$

as an analytic function in the disc $|z| < 1$. It has boundary values
$$\lim_{r \to 1} G(re^{i\theta}) = g(\theta)$$

in $L_2(S)$. Since G is an analytic function, we know from the theory of functions of a complex variable that $\log |G(re^{i\theta})|$ is subharmonic and has the mean-value property

$$\int_S \log |G(re^{i\theta})| d\theta \geq \log |G(0)| = \log |c_0|.$$

Since $G(re^{i\theta})$ has a limit in $L_2(S)$, the positive part of $\log |G|$ which is dominated by $|G|$ is uniformly integrable. For the negative part, we apply Fatou's lemma and derive our estimate.

Now for the converse. Let $f \in L_1(S)$. Assume $\int_S \log f(\theta) d\theta > -\infty$ or, equivalently, $\log f \in L_1(S)$. Define the Fourier coefficients

$$a_n = \frac{1}{4\pi} \int_S \log f(\theta) \exp[in\theta] d\theta.$$

Because $\log f$ is integrable, the $\{a_n\}$ are uniformly bounded and the power series
$$A(z) = \sum a_n z^n$$

is well-defined for $|z| < 1$. We define
$$G(z) = \exp[A(z)].$$

We will show that
$$\lim_{r \to 1} G(re^{i\theta}) = g(\theta)$$

exists in $L_2(S)$ and $f = |g|^2$, g being the boundary value of an analytic function in the disc. The integral condition on $\log f$ is then the necessary and sufficient condition for writing $f = |g|^2$ with g involving only nonnegative frequencies. Since $f(\cdot)$ is even,

$$|G(re^{i\theta})|^2 = \exp[2 \operatorname{Re} A(re^{i\theta})]$$

$$= \exp \left[2 \sum_{j=0}^{\infty} a_j r^j \cos j\theta \right]$$

$$= \exp\left[2\sum_{j=0}^{\infty} r^j \cos j\theta \left[\frac{1}{4\pi}\int_S \log f(\varphi)\cos j\varphi\, d\varphi\right]\right]$$

$$= \exp\left[\frac{1}{2\pi}\int_S \log f(\varphi)\left[\sum_{j=0}^{\infty} r^j \cos j\theta \cos j\varphi\, d\varphi\right]\right]$$

$$= \exp\left[\int_S \log f(\varphi) K(r,\theta,\varphi) d\varphi\right]$$

$$\leq \int_S f(\varphi) K(r,\theta,\varphi) d\varphi.$$

Here K is the Poisson kernel for the disc

$$K(r,\theta,\varphi) = \frac{1}{2\pi}\sum_{j=0}^{\infty} r^j \cos j\theta \cos j\varphi$$

is nonnegative and $\int_S K(r,\theta,\varphi) d\varphi = 1$. The last step is a consequence of Jensen's inequality. The function

$$f_r(\theta) = \int_S f(\varphi) K(r,\theta,\varphi) d\varphi$$

converges to f as $r \to 1$ in $L_1(S)$ by the properties of the Poisson kernel. It is, therefore, uniformly integrable. Since $|G(re^{i\theta})|^2$ is dominated by f_r, we get uniform integrability for $|G|^2$ as $r \to 1$. It is now seen that G has a limit g in $L_2(S)$ as $r \to 1$ and $f = |g|^2$. $\square$

One of the issues in the theory of time series is that of prediction. We have a stochastic process $\{X_n\}$ that we have observed for times $n \leq -1$ and we want to predict X_0. The best predictor is $E^P[X_0 \mid \mathcal{F}_{-1}]$ or, in the Gaussian linear context, it is the computation of the projection of X_0 into $\mathcal{H}_{-1}$. If we have a moving average representation, even a causal one, while it is true that X_j is spanned by $\{\xi_k : k \leq j\}$, the converse may not be true. If the two spans were the same, then the best predictor for X_0 would just be

$$\hat{X}_0 = \sum_{j \geq 1} a_j \xi_{-j},$$

obtained by dropping one term in the original representation. In fact, in answering question 2 the construction yielded a representation with this property. The quantity $|a_0|^2$ is then the prediction error. In any case, it is a lower bound.

Question 4. What is the value of prediction error and how do we actually find the predictor?

The situation is somewhat muddled. Let us assume that we have a purely nondeterministic process, i.e., a process with a spectral density satisfying $\int_S \log f(\theta) d\theta > -\infty$. Then f can be represented as

$$f = |g|^2$$

with $g \in H_2$, where by H_2 we denote the subspace of $L_2(S)$ that are boundary values of analytic functions in the disc $|z| < 1$, or, equivalently, functions $g \in L_2(S)$ with only nonnegative frequencies. For any such g, we have an analytic function
$$G(z) = G(re^{i\theta}) = \sum_{n \geq 0} a_n r^n e^{in\theta}.$$

For any choice of $g \in H_2$ with $f = |g|^2$, we have

(6.1) $$|G(0)|^2 = |a_0|^2 \leq \exp\left[\frac{1}{2\pi} \int_S \log f(\theta) d\theta\right].$$

There is a special choice of g_1 constructed in the proof of Theorem 6.12 for which

(6.2) $$|G(0)|^2 = \exp\left[\frac{1}{2\pi} \int_S \log f(\theta) d\theta\right].$$

The prediction error $\sigma^2(f)$, which depends only on f and not on the choice of g, also satisfies

(6.3) $$\sigma^2(f) \geq |G_1(0)|^2$$

for every choice of $g \in H_2$ with $f = |g|^2$. There is a choice of g_2 such that

(6.4) $$\sigma^2(f) = |G(0)|^2.$$

Therefore, from (6.1) and (6.4)

(6.5) $$\sigma^2(f) \leq \exp\left[\frac{1}{2\pi} \int_S \log f(\theta) d\theta\right].$$

On the other hand, from (6.2) and (6.3)

(6.6) $$\sigma^2(f) \geq \exp\left[\frac{1}{2\pi} \int_S \log f(\theta) d\theta\right].$$

We now have an exact formula

(6.7) $$\sigma^2(f) = \exp\left[\frac{1}{2\pi} \int_S \log f(\theta) d\theta\right]$$

for the prediction error.

As for the predictor, it is not quite that simple. In principle, it is a limit of linear combinations of $\{X_j : j \leq -1\}$ and may not always have a simple concrete representation. But we can understand it a little better. Let us consider the spaces $\mathcal{H}$ and $L_2(S; \mu)$ of square-integrable functions on S with respect to the spectral measure μ. There is a natural isomorphism between the two Hilbert spaces if we map
$$\sum a_j X_j \longleftrightarrow \sum a_j e^{ij\theta}.$$

The problem then is the question of approximating $e^{i\theta}$ in $L_2(S; \mu)$ by linear combinations of $\{e^{ij\theta} : j \leq -1\}$. We have already established that the error, which is nonzero in the purely nondeterministic case, i.e., when $d\mu = \frac{1}{2\pi} f(\theta) d\theta$ for some $f \in L_1(S)$ satisfying

$$\int_S \log f(\theta) d\theta > -\infty$$

is given by

$$\sigma^2(f) = \exp\left[\frac{1}{2\pi} \int_S \log f(\theta) d\theta\right].$$

We now want to find the best approximation.

In order to get at the predictor we have to make a very special choice of the representation $f = |g|^2$. Simply demanding $g \in L_2(S)$ will not even give causal representations. Demanding $g \in H_2$ will always give us causal representation, but there are too many of these. If we multiply $G(z)$ by an analytic function $V(z)$ that has boundary values $v(\theta)$ satisfying $|v(\theta)| = |V(e^{i\theta})| \equiv 1$ on S, then gv is another choice. If we demand that

(6.8) $$|G(0)|^2 = \exp\left[\frac{1}{2\pi} \int_S \log f(\theta) d\theta\right],$$

there is at least one choice that will satisfy it. There is still ambiguity, albeit a trivial one among these, for we can always multiply g by a complex number of modulus 1, and that will not change anything of consequence. We have the following theorem:

THEOREM 6.13 *The representation $f = |g|^2$ with $g \in H_2$ and satisfying (6.8) is unique to within a multiplicative constant of modulus 1. In other words, if $f = |g_1|^2 = |g_2|^2$ with both g_1 and g_2 satisfying (6.8), then $g_1 = \alpha g_2$ on S, where α is a complex number of modulus 1.*

PROOF: Let $F(re^{i\theta}) = \log |G(re^{i\theta})|$. It is a subharmonic function and

$$\lim_{r \to 1} F(re^{i\theta}) = \tfrac{1}{2} \log f(\theta).$$

Because

$$\lim_{r \to 1} G(re^{i\theta}) = g(\theta)$$

in $L_2(S)$, the functions are uniformly integrable in r. The positive part of the logarithm F is well controlled and therefore uniformly integrable. Fatou's lemma is applicable and we should always have

$$\limsup_{r \to 1} \frac{1}{2\pi} \int_S F(re^{i\theta}) d\theta \leq \frac{1}{4\pi} \int_S \log f(\theta) d\theta.$$

But because F is subharmonic, its average value on a circle of radius r around 0 is nondecreasing in r, and the lim sup is the same as the sup. Therefore,

$$F(0) \leq \sup_{0 \leq r < 1} \frac{1}{2\pi} \int_S F(re^{i\theta}) d\theta = \limsup_{r \to 1} \frac{1}{2\pi} \int_S F(re^{i\theta}) d\theta \leq \frac{1}{4\pi} \int_S \log f(\theta) d\theta.$$

Since we have equality at both ends, that implies a lot of things. In particular, F is harmonic and is represented via the Poisson integral in terms of its boundary value $\frac{1}{2}\log f$. In particular, G has no zeros in the disc. Obviously, F is uniquely determined by $\log f$, and by the Cauchy-Riemann equations the imaginary part of $\log G$ is determined up to an additive constant. Therefore, the only ambiguity in G is a multiplicative constant of modulus 1.

Given the process $\{X_n\}$ with trivial tail subspaces, we saw earlier that it has a representation

$$X_n = \sum_{j=0}^{\infty} a_j \xi_{n-j}$$

in terms of standard i.i.d. Gaussians, and from the construction we also know that $\xi_n \in \mathcal{H}_n$ for each n. In particular, $\xi_0 \in \mathcal{H}_0$ and can be approximated by linear combinations of $\{X_j : j \leq 0\}$. Let us suppose that $h(\theta)$ represents ξ_0 in $L_2(S; f)$. We know that $h(\theta)$ is in the linear span of $\{e^{ij\theta} : j \leq 0\}$. We want to find the function h. If $\xi_0 \longleftrightarrow h$, then by the nature of the isomorphism $\xi_n \longleftrightarrow e^{in\theta}h$, and

$$1 = \sum_{j=0}^{\infty} a_j e^{-ij\theta} h(\theta)$$

is an orthonormal expansion in $L_2(S; f)$. Also, if we let

$$G(z) = \sum_{j=0}^{\infty} a_j z^j,$$

then the boundary function $g(\theta) = \lim_{r \to 1} G(re^{i\theta})$ has the property

$$g(-\theta)h(\theta) = 1 \quad \text{and so} \quad h(\theta) = \frac{1}{g(-\theta)}.$$

Since the function G that we constructed has the property

$$|G(0)|^2 = |a_0|^2 = \sigma^2(f) = \exp\left[\frac{1}{2\pi}\int_S \log f(\theta) d\theta\right],$$

it is the canonical choice determined earlier, up to a multiplicative constant of modulus 1. The predictor, then, is clearly represented by the function

$$\hat{1}(\theta) = 1 - a_0 h(\theta) = 1 - \frac{g(0)}{g(-\theta)}.$$

□

EXAMPLE 6.2. A wide class of examples are given by densities $f(\theta)$ that are rational trigonometric polynomials of the form

$$f(\theta) = \frac{|\sum A_j e^{ij\theta}|^2}{|\sum B_j e^{ij\theta}|^2}.$$

We can always multiply by $e^{ik\theta}$ inside the absolute value and assume that
$$f(\theta) = \frac{|P(e^{i\theta})|^2}{|Q(e^{i\theta})|^2},$$
where $P(z)$ and $Q(z)$ are polynomials in the complex variable z. The symmetry of f under $\theta \to -\theta$ means that the coefficients in the polynomial have to be real. The integrability of f will force the polynomial Q not to have any zeros on the circle $|z| = 1$. Given any two complex numbers c and z such that $|z| = 1$ and $c \neq 0$,
$$|z - c| = |\bar{z} - \bar{c}| = \left|\frac{1}{z} - \bar{c}\right| = |1 - \bar{c}z| = |c|\left|z - \frac{1}{\bar{c}}\right|.$$
This means in our representation for f, first we can omit terms that involve powers of z that have only modulus 1 on S. Next, any term $(z - c)$ that contributes a nonzero root c with $|c| < 1$ can be replaced by $c(z - \frac{1}{\bar{c}})$ and thus move the root outside the disc without changing the value of f. We can therefore rewrite
$$f(\theta) = |g(\theta)|^2 \quad \text{with} \quad G(z) = \frac{P(z)}{Q(z)}$$
with new polynomials P and Q that have no roots inside the unit disc and with perhaps P alone having roots on S; clearly
$$h(\theta) = \frac{Q(e^{i\theta})}{P(e^{i\theta})}.$$
If P has no roots on S, we have a nice convergent power series for $\frac{Q}{P}$ with a radius of convergence larger than 1, and we are in a very good situation. If $P = 1$, we are in an even better situation with the predictor expressed as a finite sum. If P has a root on S, then it could be a little bit of a mess, as the next exercise shows.

EXERCISE 6.16. Assume that we have a representation of the form
$$X_n = \xi_n - \xi_{n-1}$$
in terms of standard i.i.d. Gaussians. How would you predict X_1 based on $\{X_j : j \leq 0\}$?

EXERCISE 6.17. An autoregressive scheme is a representation of the form
$$X_n = \sum_{j=1}^{k} a_j X_{n-j} + \sigma \xi_n,$$
where ξ_n is a standard Gaussian independent of $\{(X_j, \xi_j) : j \leq (n-1)\}$. In other words, the predictor
$$\hat{X}_n = \sum_{j=1}^{k} a_j X_{n-j}$$
and the prediction error σ^2 are specified for the model. Can you always find a stationary Gaussian process $\{X_n\}$ with spectral density $f(\theta)$ that is consistent with the model?

CHAPTER 7

Dynamic Programming and Filtering

7.1. Optimal Control

Optimal control or dynamic programming is a useful and important concept in the theory of Markov processes. We have a state space X and a family π_α of transition probability functions indexed by a parameter $\alpha \in \mathcal{A}$. The parameter α is called the *control parameter* and can be chosen at will from the set $\mathcal{A}$. The choice is allowed to vary over time; i.e., α_j can be the parameter of choice for the transition from x_j at time j to x_{j+1} at time $j + 1$. The choice can also depend on the information available up to that point; i.e., α_j can be an $\mathcal{F}_j$-measurable function. Then the conditional probability $P\{x_{j+1} \in A \mid \mathcal{F}_j\}$ is given by $\pi_{\alpha_j(x_0, x_1, \ldots, x_j)}(x_j, A)$. Of course, in order for things to make sense, we need to assume some measurability conditions. We have a payoff function $f(x_N)$, and the object is to maximize $E\{f(x_N)\}$ by a suitable choice of the functions $\{\alpha_j(x_0, x_1, \ldots, x_j) : 0 \leq j \leq N - 1\}$. The idea (Bellman's) of dynamic programming is to define recursively (by backward induction), for $0 \leq j \leq N - 1$, the sequence of functions

(7.1) $$V_j(x) = \sup_\alpha \int V_{j+1}(y) \pi_\alpha(x, dy)$$

with

$$V_N(x) = f(x),$$

as well as the sequence $\{\alpha_j^*(x) : 0 \leq j \leq N - 1\}$ of functions that provide the supremum in (7.1),

$$V_j(x) = \int V_{j+1}(y) \pi_{\alpha_j^*(x)}(x, dy) = \sup_\alpha \int V_{j+1}(y) \pi_\alpha(x, dy).$$

We then have

THEOREM 7.1 *If the Markov chain starts from x at time 0, then $V_0(x)$ is the best expected value of the reward. The "optimal" control is Markovian and is provided by $\{\alpha_j^*(x_j)\}$.*

PROOF: It is clear that if we pick the control as α_j^*, then we have an inhomogeneous Markov chain with transition probability

$$\pi_{j, j+1}(x, dy) = \pi_{\alpha_j^*(x)}(x, dy),$$

and if we denote by P_x^* the process corresponding to it that starts from the point x at time 0, we can establish by induction that

$$E^{P_x^*}\{f(x_N) \mid \mathcal{F}_{N-j}\} = V_{N-j}(x_{N-j})$$

for $1 \leq j \leq N$. Taking $j = N$, we obtain

$$E^{P_x^*}\{f(x_N)\} = V_0(x).$$

To show that $V_0(x)$ is optimal, for any admissible (not necessarily Markovian) choice of controls, if P is the measure on $\mathcal{F}_N$ corresponding to a starting point x,

$$E^P\{V_{j+1}(x_{j+1}) \mid \mathcal{F}_j\} \leq V_j(x_j),$$

and now it follows that

$$E^P\{f(x_N)\} \leq V_0(x).$$

□

EXERCISE 7.1. The problem could be modified by making the reward function equal to

$$E^P\left\{\sum_{j=0}^N f_j(\alpha_{j-1}, x_j)\right\}$$

and thereby incorporating the cost of control into the reward function. Work out the recursion formula for the optimal reward in this case.

7.2. Optimal Stopping

Certain optimization problems are called *optimal stopping problems*. We have a Markov chain with transition probability $\pi(x, dy)$ and time runs from 0 to N. We have the option to stop at any time based on the history up to that time. If we stop at time k in the state x, the reward is $f(k, x)$. The problem then is to maximize $E_x\{f(\tau, x_\tau)\}$ over all stopping times $0 \leq \tau \leq N$. If $V(k, x)$ is the optimal reward if the game starts from x at time k, the best we can do starting from x at time $k-1$ is to earn a reward of

$$V(k-1, x) = \max\left[f(k-1, x), \int V(k, y)\pi(x, dy)\right].$$

Starting with $V(N, x) = f(N, x)$, by backwards induction we can get $V(j, x)$ for $0 \leq j \leq N$. The optimal stopping rule is given by

$$\bar{\tau} = \{\inf k : V(k, x_k) = f(k, x_k)\}.$$

THEOREM 7.2 *For any stopping time τ with $0 \leq \tau \leq N$,*

$$E_x\{f(\tau, x_\tau)\} \leq V(0, x) \quad \text{and} \quad E_x\{f(\bar{\tau}, x_{\bar{\tau}})\} = V(0, x).$$

PROOF: Because

$$V(k, x) \geq \int V(k+1, y)\pi(x, dy),$$

7.2. OPTIMAL STOPPING

we conclude that $V(k, x_k)$ is a supermartingale and an application of Doob's stopping theorem proves the first claim. On the other hand, if $V(k, x) > f(k, x)$, we have

$$V(k, x) = \int V(k+1, y) \pi(x, dy),$$

and this means $V(\bar\tau \wedge k, x_{\bar\tau \wedge k})$ is a martingale, and this establishes the second claim. □

EXAMPLE 7.1 (The Secretary Problem). An interesting example is the following game: We have a lottery with N tickets. Each ticket has a number on it. The numbers $a_1, a_2, \ldots, a_N$ are distinct, but the player has no idea of what they are. The player draws a ticket at random and looks at the number. He can either keep the ticket or reject it. If he rejects it, he can draw another ticket from the remaining ones and again decide if he wants to keep it. The information available to him is the numbers on the tickets he has so far drawn and discarded, as well as the number on the last ticket that he has drawn and is holding. If he decides to keep the ticket at any stage, then the game ends and that is his ticket. Of course, if he continues on till the end, rejecting all of them, he is forced to keep the last one. The player wins only if the ticket he keeps is the one that has the largest number written on it. He cannot go back and claim a ticket that he has already rejected, and he cannot pick a new one unless he rejects the one he is holding. Assuming that the draws are random at each stage, how can the player maximize the probability of winning? How small is this probability?

It is clear that the strategy to pick the first or the last or any fixed draw has the probability of $\frac{1}{N}$ to win. It is not a priori clear that the probability p_N of winning under the optimal strategy remains bounded away from 0 for large N. It seems unlikely that any strategy can pick the *winner* with significant probability for large values of N. Nevertheless, the following simple strategy shows that

$$\liminf_{N \to \infty} p_N \geq \tfrac{1}{4}.$$

Let half the draws go by, no matter what, and then pick the first one which is the highest among the tickets drawn up to the time of the draw. If the second best has already been drawn and the best is still to come, this strategy will succeed. This has probability nearly $\frac{1}{4}$. In fact, the strategy works if the k best tickets have not been seen during the first half, $(k+1)^{\text{th}}$ has been, and among the k best the highest shows up first in the second half. The probability for this is about $\frac{1}{k 2^{k+1}}$, and as these are disjoint events

$$\liminf_{N \to \infty} p_N \geq \sum_{k \geq 1} \frac{1}{k 2^{k+1}} = \frac{1}{2} \log 2.$$

If we decide to look at the first Nx tickets rather than $\frac{N}{2}$, the lower bound becomes $x \log \frac{1}{x}$, and an optimization over x leads to $x = \frac{1}{e}$ and the resulting lower bound

$$\liminf_{N \to \infty} p_N \geq \frac{1}{e}.$$

We will now use the method of optimal stopping to decide on the best strategy for every N and show that the procedure we described is about the best. Since the only thing that matters is the ordering of the numbers, the numbers themselves have no meaning. Consider a Markov chain with two states 0 and 1. The player is in state 1 if he is holding the largest ticket so far. Otherwise he is in state 0. If he is in state 1 and stops at stage k, i.e., when k tickets have been drawn, the probability of his winning is easily calculated to be $\frac{k}{N}$. If he is in state 0, he has to go on, and the probability of landing on 1 at the next step is calculated to be $\frac{1}{k+1}$. If he is at 1 and decides to play on, the probability is still $\frac{1}{k+1}$ for landing on 1 at the next stage. The problem reduces to optimal stopping for a sequence $X_1, X_2, \ldots, X_N$ of independent random variables with $P\{X_i = 1\} = \frac{1}{i+1}$, $P\{X_i = 0\} = \frac{i}{i+1}$, and a reward function of $f(i, 1) = \frac{i}{N}$, $f(i, 0) = 0$.

Let us define recursively the optimal probabilities

$$V(i, 0) = \frac{1}{i+1} V(i+1, 1) + \frac{i}{i+1} V(i+1, 0)$$

and

$$V(i, 1) = \max\left[\frac{i}{N}, \frac{1}{i+1} V(i+1, 1) + \frac{i}{i+1} V(i+1, 0) \right]$$
$$= \max\left[\frac{i}{N}, V(i, 0) \right].$$

It is clear what the optimal strategy is. We should always draw if we are in state 0; i.e., we are sure to lose if we stop. If we are holding a ticket that is the largest so far, we should stop provided

$$\frac{i}{N} > V(i, 0) \quad \text{and go on if} \quad \frac{i}{N} < V(i, 0).$$

Either strategy is acceptable in case of equality. Since $V(i+1, 1) \geq V(i+1, 0)$ for all i, it follows that $V(i, 0) \geq V(i+1, 0)$. There is, therefore, a critical k ($= k_N$) such that $\frac{i}{N} \geq V(i, 0)$ if $i \geq k$ and $\frac{i}{N} \leq V(i, 0)$ if $i \leq k$. The best strategy is to wait till k tickets have been drawn, discarding every ticket, and then pick the first one that is the best so far. The last question is the determination of $k = k_N$. For $i \geq k$,

$$V(i, 0) = \frac{1}{i+1} \frac{i+1}{N} + \frac{i}{i+1} V(i+1, 0) = \frac{1}{N} + \frac{i}{i+1} V(i+1, 0)$$

or

$$\frac{V(i, 0)}{i} - \frac{V(i+1, 0)}{i+1} = \frac{1}{N} \cdot \frac{1}{i},$$

telling us

$$V(i, 0) = \frac{i}{N} \sum_{j=i}^{N-1} \frac{1}{j} \quad \text{so that} \quad k_N = \inf\left\{ i : \frac{1}{N} \sum_{j=i}^{N-1} \frac{1}{j} < \frac{1}{N} \right\}.$$

Approximately $\log N - \log k_N = 1$ or $k_N = N/e$.

7.3. Filtering

The problem in filtering is that there is an underlying stochastic process that we cannot observe. There is a related stochastic process "driven" by the first one that we can observe, and we want to use our information to make conclusions about the state of the unobserved process. A simple but extreme example is when the unobserved process does not move and remains at the same value. Then it becomes a parameter. The driven process may be a sequence of independent identically distributed random variables with densities $f(\theta, x)$ where θ is the unobserved, unchanging underlying parameter. We have a sample of n independent observations $X_1, X_2, \ldots, X_n$ from the common distribution $f(\theta, x)$, and our goal is then nothing other than parameter estimation.

We shall take a Bayesian approach. We have a prior distribution $\mu(d\theta)$ on the space of parameters Θ, and this can be modified to an a posteriori distribution after the sample is observed. We have the joint distribution

$$\prod_{i=1}^{n} f(\theta, x_i) dx_i \mu(d\theta),$$

and we calculate the conditional distribution of

$$\mu_n(d\theta \mid x_1 x_2 \cdots x_n),$$

given $x_1, x_2, \ldots, x_n$. This is our best informed guess about the nature of the unknown parameter. We can use this information as we see fit. If we have an additional observation x_{n+1}, we need not recalculate everything, but we can simply update by viewing μ_n as the new prior and calculating the posterior after a single observation x_{n+1}.

We will just work out a single illustration of this known as the Kallman-Bucy filter. Let us suppose that the unobserved process $\{x_n\}$ is a Gaussian Markov chain

$$x_{n+1} = \rho x_n + \sigma \xi_{n+1}$$

with $0 < \rho < 1$ and the noise terms $\{\xi_n\}$ are i.i.d. normally distributed random variables with mean 0 and variance 1. The observed process y_n is given by

$$y_n = x_n + \eta_n,$$

where the $\{\eta_j\}$ are again independent standard Gaussians that are independent of the $\{\xi_j\}$ as well. If we start with an initial distribution for x_0, say one that is Gaussian with mean m_0 and variance σ_0^2, we can compute the joint distribution of x_0, x_1, and y_1 and then the conditional of x_1 given y_1. This becomes the new distribution of the state x_1 based on the observation y_1. This allows us to calculate recursively at every stage.

Let us do this explicitly now. The distribution of x_1, y_1 is jointly normal with means $(\rho m_0, \rho m_0)$, variances $(\rho^2 \sigma_0^2 + \sigma^2, \rho^2 \sigma_0^2 + \sigma^2 + 1)$, and covariance $(\rho^2 \sigma_0^2 +$

σ^2). The posterior distribution of x_1 is again normal with mean

$$m_1 = \rho m_0 + \frac{(\rho^2 \sigma_0^2 + \sigma^2)}{(\rho^2 \sigma_0^2 + \sigma^2 + 1)}(y_1 - \rho m_0)$$

$$= \frac{\rho}{(\rho^2 \sigma_0^2 + \sigma^2 + 1)} m_0 + \frac{(\rho^2 \sigma_0^2 + \sigma^2)}{(\rho^2 \sigma_0^2 + \sigma^2 + 1)} y_1$$

and variance

$$\sigma_1^2 = (\rho^2 \sigma_0^2 + \sigma^2)\left(1 - \frac{(\rho^2 \sigma_0^2 + \sigma^2)}{(\rho^2 \sigma_0^2 + \sigma^2 + 1)}\right) = \frac{(\rho^2 \sigma_0^2 + \sigma^2)}{(\rho^2 \sigma_0^2 + \sigma^2 + 1)}.$$

After a long time, while the recursion for m_n remains the same,

$$m_n = \frac{\rho}{(\rho^2 \sigma_{n-1}^2 + \sigma^2 + 1)} m_{n-1} + \frac{(\rho^2 \sigma_{n-1}^2 + \sigma^2)}{(\rho^2 \sigma_{n-1}^2 + \sigma^2 + 1)} y_n,$$

the variance σ_n^2 has an asymptotic value σ_∞^2 given by the solution of

$$\sigma_\infty^2 = \frac{(\rho^2 \sigma_\infty^2 + \sigma^2)}{(\rho^2 \sigma_\infty^2 + \sigma^2 + 1)}.$$

Bibliography

[1] Ahlfors, L. V. *Complex analysis.* An introduction to the theory of analytic functions of one complex variable. Third edition. International Series in Pure and Applied Mathematics. McGraw-Hill, New York, 1978.

[2] Dym, H., and McKean, H. P. *Fourier series and integrals.* Probability and Mathematical Statistics, 14. Academic Press, New York–London, 1972.

[3] Halmos, P. R. *Measure theory.* Van Nostrand, New York, 1950.

[4] Kolmogorov, A. N. *Foundations of the theory of probability.* Translation edited by Nathan Morrison, with an added bibliography by A. T. Bharuch-Reid. Chelsea, New York, 1956.

[5] Parthasarathy, K. R. *An introduction to quantum stochastic calculus.* Monographs in Mathematics, 85. Birkhäuser, Basel, 1992.

[6] ———. *Probability measures on metric spaces.* Probability and Mathematical Statistics, 3. Academic Press, New York–London, 1967.

[7] Royden, H. L. *Real analysis.* Third edition. Macmillan, New York, 1988.

[8] Stroock, D. W., and Varadhan, S. R. S. *Multidimensional diffusion processes.* Grundlehren der Mathematischen Wissenschaften [Fundamental Principles of Mathematical Sciences], 233. Springer-Verlag, Berlin–New York, 1979.

Index

σ-field, 2

accompanying laws, 55

Bellman, 157
Berry, 69
Berry-Essen theorem, 69
binomial distribution, 19
Birkhoff, 131, 133
   ergodic theorem of, 131, 133
Bochner, 20, 30, 33, 147
   theorem of, 20, 30
      for the circle, 33
Borel, 40
Borel-Cantelli lemma, 40
bounded convergence theorem, 10
branching process, 105
Bucy, 161

Cantelli, 40
Caratheodory
   extension theorem, 4
Cauchy, 22
Cauchy distribution, 22
central limit theorem, 50
   under mixing, 146
change of variables, 13
Chapman, 85
Chapman-Kolmogorov equations, 85
characteristic function, 19
   uniqueness theorem, 21
Chebyshev, 38
Chebyshev's inequality, 38
compound Poisson distribution, 54
conditional expectation, 73, 79
   Jensen's inequality, 80
conditional probability, 73, 81
   regular version, 82
conditioning, 73
continuity theorem, 25
control, 157

convergence
   almost everywhere, 9
   in distribution, 24
   in law, 24
   in probability, 9
convolution, 36
countable additivity, 2
covariance, 17
covariance matrix, 17
Cramér, 25

degenerate distribution, 19
Dirichlet, 21
Dirichlet integral, 21
disintegration theorem, 84
distribution
   joint, 14
   of a random variable, 13
distribution function, 6
dominated convergence theorem, 12
Doob, 110, 111, 115, 118, 120
   decomposition theorem of, 115
   inequality of, 110, 111
   stopping theorem of, 118
   up-crossing inequality of, 120
double integral, 16
dynamic programming, 157

ergodic invariant measure, 135
ergodic process, 135
   extremality of, 135
ergodic theorem, 131
   almost sure, 131, 133
   maximal, 133
   mean, 131
ergodicity, 135
Esseen, 69
exit probability, 125
expectation, 17

exponential distribution, 22
    two sided, 22
extension theorem, 4

Fatou, 11
Fatou's lemma, 11
field, 2
    σ-field generated by, 3
filter, 161
finite additivity, 2
Fubini, 16
Fubini's theorem, 16

gamma distribution, 22
Gaussian distribution, 22
Gaussian process, 147
    stationary, 147
        autoregressive schemes, 155
        causal representation of, 147
        moving average representation of, 147
        prediction of, 151
        prediction error of, 151
        predictor of, 151
        rational spectral density, 155
        spectral density of, 147
        spectral measure of, 147
generating function, 23
geometric distribution, 21

Hahn, 75
Hahn-Jordan decomposition, 75

independent events, 35
independent random variables, 35
indicator function, 7
induced probability measure, 13
infinitely divisible distributions, 59
integrable functions, 11
integral, 7
invariant measures, 131
inversion theorem, 21
irrational rotations, 137

Jensen, 80
Jordan, 75

Kallman, 161
Kallman-Bucy filter, 161
Khintchine, 64
Kolmogorov, 1, 41, 43, 46, 47, 49, 85
    consistency theorem of, 41, 42
    inequality of, 43
    one series theorem of, 46
    three series theorem of, 47

two series theorem of, 46
zero-one law of, 49

Lévy, 25, 44, 61, 64
    inequality of, 44
    theorem of, 44
Lévy measures, 61
Lévy-Khintchine representation, 64
law of large numbers
    strong, 42
    weak, 37
law of the iterated logarithm, 66
Lebesgue, 5
    extension theorem, 5
Lindeberg, 51, 53
    condition of, 51
    theorem of, 51
Lipschitz, 78
Lipschitz condition, 78
Lyapunov, 53
    condition of, 53

mapping, 13
Markov, 85
    chain, 85
    process, 85
        homogeneous, 85
Markov chain
    aperiodic, 98
    invariant distribution for, 89
    irreducible, 91
    periodic behavior, 98
    stationary distribution for, 89
Markov process
    invariant measures
        ergodicity, 138
    invariant measures for, 138
    mixing, 141
    reversible, 138
    stationary, 138
Markov property, 87
    strong, 90
martingale difference, 109
martingale transform, 121
martingales, 109
    almost sure convergence of, 113, 116
    central limit theorem for, 144
    convergence theorem, 112
    sub-, 110
    super-, 110
maximal ergodic inequality, 134
mean, 17
measurable function, 7

INDEX

measurable space, 13
moments, 20, 23
   generating function, 23
   uniqueness from, 23
monotone class, 2, 4
monotone convergence theorem, 11

negative binomial distribution, 22
Nikodym, 76
normal distribution, 22

optimal control, 157
optimal stopping, 158
option pricing, 122
optional stopping theorem, 118
Ornstein, 142
Ornstein-Uhlenbeck process, 142
outer measure, 4

Poisson, 21, 54, 151
Poisson distribution, 21
Poisson kernel, 151
positive definite function, 19, 30
probability space, 6
product σ-field, 15
product measure, 14
product space, 14

queues, 100

Radon, 76
Radon-Nikodym
   derivative, 76
   theorem, 76
random variable, 7
random walk, 88
   recurrence, 128
   simple, 98
   transience, 127
recurrence, 91
   null, 91
   positive, 91
recurrent states, 97
renewal theorem, 94
repeated integral, 16
Riemann-Stieltjes integral, 18

secretary problem, 159
signed measure, 75
simple function, 7
Stirling, 39
Stirling's formula, 39, 50
stochastic matrix, 90
stopped σ-field, 118

stopping time, 89, 117

transformations, 13
   measurable, 13
   measure preserving, 131
      isometries from, 131
transience, 91
transient states, 97
transition operator, 124
transition probability, 85
   stationary, 85
Tulcea, 85
   theorem of, 85

Uhlenbeck, 142
uniform distribution, 22
uniform infinitesimality, 54
uniform tightness, 29
up-crossing inequality, 120
urn problem, 103

variance, 17

weak convergence, 24
Weierstrass, 24
   factorization, 24